Managerial Operations Research

WILLIAM D. BRINCKLOE
Professor of Management Science
Graduate School of Public and International Affairs
University of Pittsburgh

McGraw-Hill Book Company
New York   St. Louis   San Francisco   London
Sydney   Toronto   Mexico   Panama

MANAGERIAL OPERATIONS RESEARCH

Copyright © 1969 by McGraw-Hill, Inc. All Rights Reserved. Printed in the United States of America. No part of this publication may be reproduced, stored in a retrieval system, or transmitted, in any form or by any means, electronic, mechanical, photocopying, recording, or otherwise, without the prior written permission of the publisher.

Library of Congress Catalog Card Number 77-77556

07903

1234567890 MAMM 754321069

# Managerial Operations Research

Preface

This book is for today's versatile manager, whose expanding horizons require him to understand the management uses of operations research.

This executive need not be an expert in quantitative methods, but he operates in a complex environment where they are becoming almost commonplace, and he cannot leave the field to experts. He must know when quantitative methods are valuable and when they are reliable. He must have the confidence and the feel to accept or override the recommendations of the operations research staff as the situation demands. Just as a modern executive needs a comprehension of accounting or law, so he must appreciate the nature and role of corporate systems analysis.

Staff operations researchers are not immune from the occupational failing Lord Salisbury noted in specialists: the physician finds all things unhealthy, the priest finds all things impious, and other specialists seek answers to general problems within their special fields. When the United States Navy first moved from sail to steam, the new chief engineers advanced the thesis that "steam enginery" was too complex for sailing-ship captains to comprehend—and the latter almost abdicated control before they realized that steam was simply another tool for the seagoing executive to use.

So it is with operations research. Whatever the manager's field—

finance, manufacturing, marketing, research, distribution, purchasing, warehousing, strategic planning—the broadening scope and accelerating pace of corporate life are forcing him to find more powerful methods for solving difficult management problems. Whether he is experienced in using quantitative techniques, or is plunging into these waters cold, he needs to enlarge his understanding of how to structure a large or complex task so that he can solve it.

This book presents a wide range of quantitative management techniques in easily understood style, and aims at practical problems that the real-life executive encounters every day. It should be read by practicing managers whose problems beset them now; and it should be studied by business students who are not majoring in operations research but recognize their need for a clear grasp of the rationale behind these powerful methods.

There are some necessary acknowledgments. The tedious typing of the manuscript was by courtesy of my untiring secretary, Roxanne Campana. The critical path schedule for installation of a gas-cracking furnace shown in Table 8-2 is reproduced by courtesy of Walter Cosinuke of the Catalytic Cracking Company. The waiting-line delay function shown in Fig. 9-4 is reproduced by courtesy of E. Duckworth and University Paperbacks of London, England. The Simplex Tableau shown in Fig. 10-4 was developed first by George B. Dantzig. The input/output analysis described in Chap. 11 was developed first by Wassily Leontief. The incremental analysis and expected value concepts in Chap. 5 were developed first by Robert Schlaifer.

And I must make a final grateful acknowledgment to the individual whose unfailing encouragement provided the spur that got this book written at last: my wife.

William D. Brinckloe

Contents

	Preface	v
Chapter 1	WHAT IS MANAGERIAL OPERATIONS RESEARCH?	1
Chapter 2	STATISTICAL UNDERPINNINGS OF THE NEW MANAGEMENT	15
Chapter 3	BUSINESS STRATEGIES FOR INCREASED PROFIT	39
Chapter 4	BUSINESS CLIMATE AND PROFIT	57
Chapter 5	PROFITABLE INVENTORY MANAGEMENT	68
Chapter 6	CAPITAL EXPENDITURES	92
Chapter 7	MARKET STUDIES AND LOCATIONAL ANALYSIS	112
Chapter 8	NETWORK MANAGEMENT	123
Chapter 9	WAITING LINES AND SERVICE TIMES	150
Chapter 10	PROGRAMMING	173
Chapter 11	INPUT/OUTPUT ANALYSIS	205
Epilogue	THE FUTURE	226
	Index	229

Managerial
Operations
Research

What Is Managerial Operations Research?

> In the competitive executive market in which he operates today, the highly motivated manager realizes that his progress through the executive hierarchy will depend importantly on his ability to supplement personal skills with the best of the new managerial tools.
>
> *Editors of Harvard Business Review*

The practicing manager at which this book aims knows his job well—or at least he thought he did. Lately, though, he has been getting a little uneasy. He reads of managerial revolutions and computerized information systems, and he hears predictions that those who have not seen the light are doomed to failure. The chilling automation study of one foundation predicts that growing numbers of managers will find themselves displaced, and he hopes they don't mean him. He overhears the easy jargon of those apparently in rapport with the computers, and wonders if he is somehow out of tune. He worries that he is missing the boat, and wishes he knew what there was in all this for him, but he doesn't know where to turn.

He *is* missing the boat. There is a great deal in all this for him. In the last few years, stimulated by the applicability of simple mathematics to wartime military problems, quantitative decision-making has invaded the executive suite. The simultaneous arrival of the digital computer, whose speed and growing simplicity of application make possible all sorts of helpful investigations, has accelerated this trend. Every executive want ad shows more corporations setting up operations research groups, and the hints are clear that line executives are more acceptable with a little quantitative know-how.

The aim of this book is to provide the practicing line executive with an understanding of operations research and related techniques that have become essential tools of modern business and industrial management.

OPERATIONS RESEARCH: A DEFINITION

Operations research is one name given to the broad and rather amorphous science of establishing mathematical or other explicit relationships that describe the key elements of some actual physical or administrative process with reasonable fidelity, and of drawing useful conclusions about the actual process through analysis of these relationships. (There are other names which may mean more or less the same thing: systems analysis, operations analysis, management science, quantitative method, and so on; sometimes, though, these terms have quite a different meaning, as when the computer programmer uses "systems analysis" to mean arranging a process into a sequence of steps so it can be accomplished on the computer.) The essence of operations research is that it puts aside the less significant factors and describes the way in which the key factors contribute to the key product.

A car dealer appraising a trade-in ignores hundreds of trivial factors —color, miles driven, condition of engine, type of use, former owner, make of tires, and so on—and makes his estimate in minutes on the basis of model, age, and appearance. He has learned that if he took days to examine each car minutely, in the long run the results would be very close to his on-the-spot judgment. Bank loan officers have devised the same short-cut method for loan approval, to the point where many banks turn the screening task over to computers.* Both of these procedures are representative of operations research techniques, in that they seek to peer through the thicket of facts and see the camouflaged trail through the forest; the operations research specialist is not needed, however, unless the path to a solution is less obvious than in the used-car case.

THE BACKGROUND OF OPERATIONS RESEARCH

Many scholars (and gamblers) have contributed to the underpinnings of operations research through the years, but the father of quantitative

---

* See J. A. Vaughan and A. M. Porat, Managerial Reactions to Computers, *J. Am. Bankers Assoc.*, April, 1967.

WHAT IS MANAGERIAL OPERATIONS RESEARCH?

decision-making may be Frederick William Lanchester,* whose celebrated "N-square law" assessed the fighting power of opposing forces. When cavemen fought club against club, he reasoned, tactical concentration was precluded by the limited range of weapons; thus a Donnybrook between 1,000 "Red" and 1,400 "Blue" cavemen would end up with everyone dead except 400 Blues. When weapon range permitted tactical concentration (10 riflemen could concentrate on a single enemy), the *rate* of losses became proportional to the opponent's force—a rate that constantly increased for the weaker, as the stronger's relative advantage kept growing. The solution of Lanchester's resulting differential equations suggested that forces of equal man-for-man fighting power have overall strength proportional to the *squares* of their respective numerical strengths. Let the 1,000 Reds and 1,400 Blues be riflemen rather than cavemen, and since the square of 1,400 is approximately twice the square of 1,000, the Reds are annihilated with 1,000 victorious Blues still on the field.

Lanchester tested his theory against Admiral Nelson's Plan of Battle at Trafalgar. Nelson, calculating that the French would have 46 ships against his 40, divided his force into a main body of 32 ships to engage 23 ships in the French rear, and a second body of 8 ships to hold off the 23 downwind Frenchmen in the van until the rear 23 ships were defeated and his squadrons could join forces. Lanchester calculated Nelson's N-square strategy as follows:

French fighting strength	*English fighting strength*
$(23)^2 = 529$	$(32)^2 = 1,024$
$(23)^2 = 529$	$(8)^2 = 64$
$\overline{1,058}$	$\overline{1,088}$

English margin of superiority = $\sqrt{1,088 - 1,058} = \sqrt{30} = 5\frac{1}{2}$ ships.

Nelson's strategy was exactly optimum according to Lanchester's theory; if he had divided his force any more evenly, say 31 and 9, his resultant strength would have been less than that of the French (only 1,042 versus 1,058), and if he had divided more unevenly, say 33 and 7, the weaker force might not have been able to carry out its initial assignment of holding off the French van without a battle. Perhaps this proves Nelson right because his strategy corresponds to the mathematical optimum; perhaps it proves Lanchester right because his theory confirms

---

* F. W. Lanchester, *Aircraft in Warfare: The Dawn of the Fourth Arm*, Constable & Co., Ltd., London, 1916.

a professional's decision. The point is that Lanchester attempted to make explicit the underlying factors determining strategic choice, tested his "mathematical model" in a specific real-world case, and then offered it for general application to future cases. Ever since Lanchester, operations researchers have been trying to do precisely the same thing.

THE EVOLUTION OF OPERATIONS RESEARCH

The first application of our operations research concepts in this country appeared in the mid-1930s (although the term was not used at that time), when Congress was trying to assess the value to the nation of various river, harbor, and irrigation projects. Faced with the fact that there was no measurable payoff from these public projects, the government agencies and Congressional committees worked up techniques for estimating the equivalent social payoff. Irrigation water goes to the farmer free, but its imputed value is measured by estimating what the farmer would be willing to pay if it were not free. These techniques were expanded to include the side benefits of such projects: in addition to the navigational benefits of a canal system, what are the recreational benefits from any lakes thus created? Although these projects often were scorned as pork-barrel measures, the cost-effectiveness techniques they spawned foreshadowed the systems analysis developed in the Pentagon under Robert McNamara, and the program budgeting system now spreading to all government departments.

Operations research demonstrated its value on a large scale in World War II, as commanders needed help to comprehend the implications of intelligence collected over large theaters of operation. Some of its early triumphs were dramatic, partly because many of these unfamiliar operations provided such fertile fields for exploitation. When aircraft depth charges met with poor success in sinking submarines, analysis showed that their setting presupposed too fast a diving rate for submarines surprised on the surface, so that they were exploding too deep; halving the depth setting increased the kill rate dramatically with no increase whatsoever in resources allocated. Study of convoy operations showed that wolf-pack U-boats sank a relatively fixed number of freighters per attack, almost independently of convoy size; the resultant shift to fewer but larger convoys drastically reduced overall losses and marked the turning point in the Battle of the Atlantic.*

---

* An excellent review of the wartime operations research work done by the United States Navy is contained in P. M. Morse and G. E. Kimball, *Methods of Operations Research*, John Wiley & Sons, Inc., New York, 1951.

This wartime organization extended its usefulness in the Korean War, by which time the military commander had learned to work smoothly with the analyst assigned to his command. In planning an air strike on rail lines, for example, the analyst used data on bombing and navigational accuracy, effective damage radius of different bombs, probable enemy opposition, construction details of rail embankments, etc., to make the most effective choice of various attack modes utilizing equivalent resources (such as one 1,000-pound bomb versus ten 100-pound bombs). In the Vietnamese conflict these techniques extended to broad logistics analyses of a total theater such as simulating battlefield resupply of expendables and repair parts, using alternate transportation networks involving many routes with different capacities and ambushing probabilities.

After the war, business began to adapt these military innovations to its own operations. A well-publicized statistical-analysis group from the Air Force moved as a body to the Ford Motor Company; known as the "whiz kids," its members rose to top-management spots at Ford and later in other corporations. Initial business applications were transfers from military prototypes, such as finding the most effective combination of warehouses and shipping routes to minimize a corporation's overall cost of storing and shipping, or determining the cheapest combination of input factors to produce a required product meeting certain specifications. As business gained more experience with these techniques, it developed methods directly adapted to its own needs, such as analysis of marketing strategies for maximum profit, capital investment theory, and input/output analysis for industrial planning.

A formidable business contender today is the aerospace conglomerate, which in a sense has had the best of both worlds. Much of its research has been government-supported (and in exotic areas which have attracted some of the most capable scientists), it has been encouraged and often virtually forced to use advanced techniques in conducting analysis of weapons systems and similar projects, and it has lived in an environment of lively innovation in which the operations researcher's contributions have been accepted routinely.

These companies are expanding, by merger or broadening of product lines, into traditional consumer goods areas (plus the unfolding public sector), and they can be expected to make full use of quantitative management techniques in battling it out with the companies who hold the present markets.

Today few businesses of any size overlook at least some of the advantages of managerial operations research. A recent survey of such techniques, as used by management consultants of an accounting firm that deals with leading corporations, casts an interesting light on the relative popularity of the various areas. The following shows the extent to which recent consultant reports call on each of these techniques.*

General advice	18%
Critical path	16%
Inventory theory	14%
General statistical	11%
Linear programming	10%
Other forecasting techniques	7%
Simulation	5%
Methods and procedures	4%
Computer technology	4%
Decision theory	3%
Statistical regression	2%
Exponential smoothing	1%

THE BASIC CONCEPTS OF OPERATIONS RESEARCH

Specialists contribute two things: special information and special techniques. If you must know about tax law or tar sands, computer programming or colloidal chemistry, you need *information* from a professional in the field. But if a *technique* works in businesses or departments with problems like yours, just knowing this is half the battle; and where the technique is simple, you may be able to apply it yourself. If it is not this simple (and usually it will not be), you can consider which of several approaches fits the special characteristics of your problem, and call in the right specialist for the specific problem.

Quantitative methods work, not because numbers are smarter than you are, but because they can extract from past information about your operation (or another very like yours) the maximum guidance for your future actions. Some things have worked well in the past and some have not; barring changes in outside factors, this pattern ought to continue. Even if outside factors do change, say in a way that raises or lowers gross sales, their effect is merely superimposed on the pattern of what has been happening before. The contribution of quantitative methods is their ability to separate these effects and

* Leland A. Moody of Arthur Andersen & Company, *Development of Management Sciences in the Public Accounting Profession,* presented at the 1966 meeting of Institute of Management Sciences.

associate them with their respective causes—to extract the hard facts hidden in disordered data.

Operations research draws on some basic relationships developed by economists, of which the following are representative:

The "marginal-analysis" concept provides to several claimants that last increment of resources which will bring to each the same unit measure of satisfaction. The family budgeteer, giving daughter $2 for a folk record, Junior 10 cents for bubble gum, and Mother $49.95 for a dress, hopes each purchase will bring equal pleasure per dollar. The company directors, apportioning capital investment funds to profit centers so as to produce equal return, are playing the same game.

The "standard-gamble" concept asks the decision-maker to evaluate each of his enterprises, with its estimated chance of payoff and the estimated profit or loss, against the certainty of some yardstick profit. "Would you rather spend $5,000 to bid a contract that would bring you $20,000 profit, and on which you have some specified chance of getting the award; or do business as usual with a 50-50 chance of making $6,000 or $7,000 for the same period?" If you can make realistic estimates of expected profits and associated odds, which is not easy, you have a systematic basis for an optimum choice.

Econometrics is a methodical attempt to weigh each payoff in proportion to its importance, so that you can evaluate an outcome in an important area as worth appropriately more than a similar outcome in a less important area. An econometrician would say that a manager who increases his supervisory ratio has decided that overhead deserves a higher coefficient in his profit equation.

The economist's "indifference curve" is an analytical concept whereby he can show you at what point in the evening you'd be just willing to forego that last drink in favor of a bus ticket home. (Any economics text will discuss indifference curves in the abstract, but few tell you how to construct one.)

THE ELEMENTS OF AN OPERATIONS RESEARCH ANALYSIS

Operations research has a host of applications, depending largely on what sort of person is using it. The administrator might employ it as a technique for applying the methods of science to administration, so that cause and effect can be predicted in the administrative sphere with something approaching the confidence with which this can be done in such "hard" sciences as physics or chemistry. Perhaps OR is understood best by describing the steps that go into the quantitative solution of a problem:

Determine your objective. Usually this is something more complex than "make the maximum profit." You operate under many constraints: working safely, maintaining your corporate reputation, staying within the law, and so on. Moreover, you won't elect to make the maximum profit in a specific case if the consequent course of action runs counter to a broader criterion in the long run—by neglecting maintenance or training, for example, or by concentrating on a few high-profit lines that may leave you dangerously undiversified in case of a shift in buying habits. But if you are to make your analysis quantitative, you can't translate all these qualifications and constraints into numbers; you have to select relatively simple objectives and live with them.

Determine a performance measure. If "maximum net profit" is your objective, the performance measure is simple: dollars of net profit. If your objective is more complex, a numerical measure of good performance is a little tougher. Nonetheless, you must pick one. And you must remember that your analysis really isn't measuring your achievement against your *objective*, but only against the performance measure you picked as proxy for the former. For example, if you are superintendent of a medical unit, and your objective is good performance on the part of your doctors, your *performance measure* may be the number of patients seen per day or something of the sort—which clearly isn't a foolproof proxy for what you really want, but may be the best you can devise.

Build a "mathematical model." This ominous-sounding device is nothing but a mathematical formula—as simple as you can make it—which contains the important variables or their limiting "parameters" and whose solution is the performance measure in numerical form. If your aim is net profits (P) and your two variables are number of items sold (n) and unit profit per item (p), then your mathematical model is nothing more than:

$$P = np$$

Note that, in building the model, you make use of theory insofar as you know it. In this example, the theory is obvious. In a more realistic case, you may have to do a great deal of hard thinking to decide how your parameters may be related. Furthermore, you will have to determine the coefficients that indicate how much each variable contributes to the performance measure.

It is well to become comfortable with problem-solving by means of models, for they are tremendous aids. We meet all sorts of models. A wind tunnel is a physical model; so is the wear simulator in a testing

WHAT IS MANAGERIAL OPERATIONS RESEARCH?

laboratory. The expression "He's a tiger!" is a symbolic model. An approximate mathematical model is the typical formula for current worth of an industrial building:

$$\text{Current worth} = [(\text{cost}) - (2\%\ \text{cost} \times \text{age})] \times \frac{\text{current price index}}{\text{original price index}}$$

Once you have decided, from evaluating many buildings, that the value of a building so many years old seems to correspond rather closely to this simple formula, it is economical to abandon the time-consuming estimates and depend thereafter on the formula—the model. Similarly, once you find that the arrival times and service times of customers can be represented by a model—slightly more complex but perfectly direct and understandable—you can use the model to simulate a wide variety of different procedures in a way that would be chaotic if you tried it on your real system. If you have seven repair centers, and from their operation you have constructed a mathematical model, you can operate your model to simulate nine repair centers, or five, or any number you want. And you can test the exact effect of each variable (average service or arrival time, number of servers, etc.).

THE ROLE OF THE EXECUTIVE

Never in history has the executive's task been tougher than it is today. Far from being displaced by supercomputers, the manager has been given more problems to solve. The burgeoning flow of management data has created pressures on management for action in new areas to exploit this data. "Computers can lessen the task of information-gathering," says an auto maker, "but they only increase the task of evaluation and recommendation."

It is surprising, in the face of these growing burdens, that much management makes so little use of quantitative managerial techniques. Its failure to exploit its own priceless operating statistics is astonishing. It guesses at developments when it could predict. Its "calculated risks" often are nothing of the sort, in the sense that there has been any analytical assessment of the probabilities and costs associated with various outcomes. It has fundamental misconceptions about samples or surveys —sometimes overestimating the validity of sample information, but more often buying too much data because it mistrusts conclusions inferred from samples of modest size. It does not appreciate how experience in one type of operation can be incorporated into some analytical structure that will provide synthetic experience about untried types of operation. If a number of factors are affecting its performance, it has little notion how their individual effects can be isolated and examined. It says "my tasks are different," ignoring the large core of regularity which makes prediction possible and can do away with the necessity for solving the same problems over and over again.

Why?

There are two reasons: communications and credibility. The line manager and the operations research specialist may speak such different languages that one really doesn't know what the other is saying; and often the mathematical approach seems to require such artificially limited assumptions as to be totally divorced from the untidy confusion of the real world. The manager feels that the operations research man knows very little about the totality of running a business, which is true; but he concludes therefore that the latter can be of no assistance, which is not true. The manager may accept assistance unhesitatingly in clearly defined specialist areas such as engineering or industrial hygiene, but feel that to call on a specialist in what apparently is the whole field of management smacks of abdicating the precise responsibilities for which he was hired.

Communications are hampered by the fact that you and the management scientist are different types of people. Your forte is making decisions in a hurry when you must; he might object on the grounds that there are insufficient facts for a good decision. You are intuitive, authoritative, inclined to weigh the human values; he is systematic, conservative, prone to dismiss personal factors as unmeasurable. You draw heavily on experience (even when it applies imperfectly) and instinctively weigh a fistful of imponderables to reach a swift decision; he leans on verifiable and repeatable factors, and hesitates to move before all the evidence is in—which it never is.

You don't have time to learn the mathematical foundations of operations research. If you had, you would know that when the OR man makes a statement he is not expressing his opinion but interpreting what the numerical indicators of your business are saying. His unsupported opinion of the action you should take is of little value, and you would be right to disregard it; his mathematical manipulations of data are (if he is competent) as valid as the information itself, and you should disregard it only if you consider the data suspect.

A good operations research specialist should not talk jargon. He should explain his assumptions and methods in reasonably understandable language, and if what he says sounds nonsensical, it probably is. But you must meet him halfway. You should have a notion what tools are in his kit, what sort of problem he can solve, what he needs to work with, and what are the general limitations of his methods. You should know when it would be helpful to call on him and when you ought to go it alone.

Indeed, you must meet him far more than halfway, if you are to take maximum advantage of quantitative management techniques. The operations research specialist is a staff man, who simply recommends; but yours is the responsibility for keeping the operation going, following up on decisions, and producing results in a highly competitive milieu. If you fail to call him in when you need him, or if you ask him to do the wrong things, you pay the price, not he. As each new method or device has entered the business world over the years, the executive has learned to understand it and turn it to advantage; when he has not, he has misused it and the business has suffered.* Quantitative management tools are so fundamentally intertwined with management itself that they must become basic to your managerial expertise.

You must learn to deal with measurement in management—not simply collecting data for decisions, but knowing how precise the numbers are, and how much they can be trusted. You must appreciate the value of complete and timely information about your operations, and how to collect it. You must have a feel for the underlying structure of a problem, and you must know the techniques for peeling away the nonessentials to lay bare the key variables that make the real difference. This is not to say that you must perform your own multiple regression analysis, or find your way through an elaborate linear programming problem, but you need to know the uses and limitations of such techniques —and to do this you must have some comprehension of the theories that underlie them. If you do not, you will not know how to steer the specialist to the relevant facts, how and when to incorporate your executive judgment, or how to understand and use his recommendations.

THE ROLE OF THE SPECIALIST

Where will the use of an operations research specialist pay off? Essentially the specialist deals with information that comes in (or can be put into) numerical form, and his particular skill lies in making recognizable order out of apparent disorder. Note that the disorder is only apparent —underneath the confusion there must be some systematic relationship or pattern which his quantitative methods can extract. Note, too, that some such information must exist, either in your operating statistics or in those of some other organization, which he can transfer to your problem; he cannot invent statistics.

If the disorder is almost total (if there is no underlying order to

---

* Companies who introduced computers without top-level understanding of their function, hoping to delegate the problem to specialists, are testimony to the folly and waste attending such an approach.

extract), his methods will not work and he probably cannot help; your managerial judgment will do a better job unaided. If there is little or no disorder (the information and its implications are clear-cut), his methods are not needed; you can see your way to the goal without specialist help. If his findings won't make any difference, it is foolish to use him. In a celebrated quantitative study, the analyst worked nearly two years to find out whether a price war with a competing product was desirable and to what extent; when he presented his findings to the board he learned that they wouldn't countenance a price war in any case, because it did not fit their corporate image.

You should not use the specialist unless you understand what you are asking him to explore, generally how he will do it, and how you will implement his possible recommendations. If one of his recommendations might be to undertake a major capital improvement program, either be prepared at the start to go after the funds for such a program or exclude it at the start from his list of alternatives. On the other hand, don't be too quick to tell him that certain alternatives are out of bounds; you may want to know the true benefits of a course of action, even though your present position is to exclude it from consideration, because the facts might change your mind.

Try not to let the specialist get too clear a picture of your hopes about the outcome. He is only human, and it is very tempting to produce and justify a recommendation known in advance to be highly acceptable. Remind yourself that you hope for such an outcome because of your present beliefs about the situation, and his findings may change some of those beliefs.

Steer him to the right sources of information, and give him all possible information on the relative reliability of such sources. If he thinks only a certain sort of information is available, he will extract the best conclusions from that much information; but if more is available, he can extract better information.

Give him all possible guidance on assumptions. The mathematical model he may build might strive to maximize some payoff for your organization; he may build such a model because he assumes that such a payoff constitutes the objective of your organization, but he may be very wrong. A new venture may be designed to penetrate a certain market rather than to turn maximum profit, or to achieve a certain quality level subject to attainment of some minimum profit, or to provide a training ground for budding executives even at some specified operating loss; he needs to be told this and not be made to guess, for differing goals lead to very different recommended actions.

Tell him the scope of the study. The operations researcher may not want to do a half-baked job, but he can accept close deadlines or fund limitations just as others can. If he knows you want the best conclusions he can reach by noon tomorrow, that's what you'll get—along with a statement of their shortcomings.

And remember, finally, that he is staff and you are line. His recommendations apply to that part of the problem amenable to quantitative analysis, but your judgment and your final decision apply to all of it. His findings arise out of the assumptions and data he used, which may constitute quite a limited view of the whole spectrum which you must consider. If he did a locational analysis for you, his estimate of annual volume would apply only if his assumed factors eventuated—if competing operations developed only where he said they might, if the future economic climate was as he postulated it, and if all the other intangibles followed his assumptions. If he did a good job, his assumptions are coupled with reasons why they should be legitimate, but you have yourself to blame if you slough over that part and imagine that he was predicting annual volume *come what may.*

It probably is a good deal more fun to be the manager than the specialist, but part of the fun lies in knowing your stuff—knowing enough about the specialist's work to second-guess him a bit and to bring out his very best performance. To be a generalist means to know something about everything—a tough assignment, but a challenging one.

NEEDS FULFILLED BY THIS BOOK

This book is about numbers in management, in the sense that it attempts to marshal in numerical form the business facts and interrelationships that are the raw material of managerial decision-making. It tries to bring the term "mathematical model" to life, and show how understandable and useful models can be. It is not a camouflaged course in statistics,* but a book about those managerial problems with which quantitative methods can help.

Chapter 2 discusses the basic principles of collecting facts for decision-making: how much information you need, how to interpret facts, how to combine facts and judgment, and pitfalls in collecting facts. This is an important chapter, for the principles it sets forth apply throughout the book.

Chapter 3 describes the analysis of business operations to identify strategies that produce net profit and to assess their relative weight.

* However, Stanislav Menshikov, top Soviet delegate to the 1966 Management Session at Rotterdam, says, "Managers now must know such things as mathematical and statistical analysis as well as management techniques."

Chapter 4 extends this analysis to include allowance for external economic conditions.

Chapter 5 is an introduction to inventory theory. It starts with the basic problem of finding the optimum inventory level in the face of costs that rise with increasing inventories and those that fall with increasing inventories, and ends by analyzing the situations that arise when both supply and demand are uncertain.

Chapter 6 is a discussion of capital investment policy: profitability analysis for multiproduct operations, a checklist for appraising capital projects, an assessment of various evaluation methods, and the treatment of uncertainty. It develops replacement policy guidelines for facilities subject to deteriorating utility and mounting maintenance costs, including analysis of equipment subject to sudden failure.

Chapter 7 deals with the marketing aspects of locating a business: how to assess the profitability of a location, how to determine the relative importance of various positive and negative factors, and how to predict the strength of the market.

Chapter 8 explains the concept of PERT and other scheduling techniques, and describes how they provide invaluable assistance to the manager of a complex project.

Chapter 9 explores some of the little-understood idiosyncrasies of waiting lines: factors that cause lines to grow, how to assess the optimum size of service facilities, how to predict delays and keep them within acceptable bounds. This chapter describes the useful managerial tool of simulation, which makes it possible to duplicate an administrative situation on paper and see how it behaves dynamically over time.

Chapter 10 discusses linear programming and some of its sister techniques: methods for determining the optimum resource allocation in complex systems involving multiple inputs, methods, and outputs (such as the plant assignment and routing problem for a company with many products, plants, warehouses, and markets).

The final chapter describes input/output (or interindustry analysis), a method for predicting and analyzing the demand for products or services that takes into account the complex interplay of all interrelated industries. Originally developed solely as a national planning tool, input/output techniques have been applied at a regional and even a single-company level; they permit market prediction or industrial analysis of a region with a precision and completeness not heretofore attainable.

2
Statistical Underpinnings of the New Management

INTRODUCTION

Large decisions from few facts

The essence of executive decision-making lies in using bits and pieces of information to reach conclusions far outreaching the data. You want comprehensive facts on which to base every decision, but seldom are you so fortunate. The older your tenure in the managerial saddle, the more resigned you are to this frustrating situation. You may even pride yourself on it, for if all the facts were on parade the job could be run by a lesser man.

There isn't much danger. Managers have such an insatiable demand for information that they sorely need any weapons that bear on the task of improving its flow. This chapter deals with methods of collecting facts, in small quantities, for more effective decision-making.

The economics of information

When you decide without any facts, you simply guess. The first bit of data you collect is the most valuable, for it brings you so much closer to an optimal decision than if you had no data at all. Each successive bit of data tends to be progressively less useful, contributing smaller and smaller refinements to your decision. The moment you have enough data for an optimal choice, additional facts are valueless.*

The cost of data goes the other way. The first bit is the cheapest because any facts you can gather will help. As your information store grows, you probe successively remoter crannies, often duplicating earlier data. A market study first seeks the preferences of the general public—an easy target to find; but later stages may have to find women over 40 with grammar school educations and incomes under $4,000—a tough set of privacy roadblocks to surmount.

It is impossible to set any precise break-even point, but no information search should start without some analysis of its optimal scope. You have to decide, in the light of data-collection costs, how much precision and confidence you need; then you plan to collect just enough data to satisfy your needs.

Contents of this chapter

This chapter could be a book in itself, on the subject of drawing conclusions about large populations by means of strategically selected samples. The populations may be sets of future events; the samples necessarily are in the past. The populations may be huge, or even infinite (how all possible customers may feel at all possible times about potential actions of yours, for example); cost limitations require that the samples be small. A member of the population represents only itself; each member of the sample may have to speak with acceptable accuracy for hundreds or thousands of others†—and, if selected properly, it can.

This chapter will discuss the techniques and pitfalls in sampling for decisions, will set forth the rules for striking a balance between so few facts that you are misled and so many that you're wasting money, and will show how information contains within itself a measure of its dependability.

THE IDEA BEHIND SAMPLING

The normal probability distribution

Pinballs and Probability

If your son needs a science project, get him a sheet of plywood, 10 pegs, 5 teacups, a bag of steel balls, and a funnel. Have him put the board

---

* It is as if you were deciding how to dress on a rainy morning; the first glance out the window dictates a raincoat, to which more detailed scrutiny may add overshoes; after this, even precise pressure and humidity readings contribute very little to the adequacy of your costume.

† Occasional attacks on television viewer surveys miss the mark here; it isn't the use of Mrs. Jones to represent 15,000 viewers that makes such surveys doubtful, but how Mrs. Jones was selected for this task.

STATISTICAL UNDERPINNINGS OF THE NEW MANAGEMENT   17

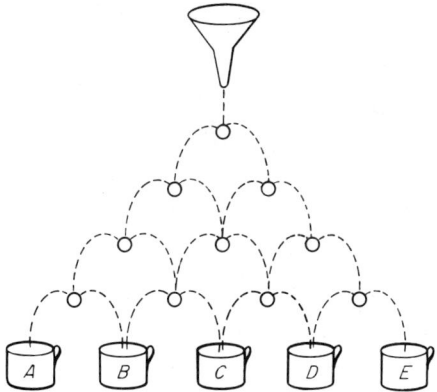

Fig. 2-1

on the wall, insert the pegs in the form of a pyramid, secure the funnel at the top so balls will drop from it onto the top peg, and arrange the cups along the bottom to catch the balls after they finish bouncing from peg to peg. If he's careful, the process will look like Fig. 2-1.

A ball landing on a peg has an equal chance of bouncing either way, so that any of the 16 paths through the maze has an equal chance of being taken. There is only 1 path to cups A and E whereas there are 4 paths to cups B and D, and 6 paths to cup C. If you drop thousands of balls, the numbers in the cups will approach the proportions $1:4:6:4:1$.

If you add another row with 5 pegs, and 6 cups to catch the balls, the expected proportions will be $1:5:10:10:5:1$. If you keep adding pins and cups, the proportions in the cups will approach the values shown in Fig. 2-2.*

Look at Fig. 2-3, the 9-cup case. Out of 256 pinballs, you would expect about 70 to drop in the middle cup and 56 more in the cup to either side; thus a total of 182, or 71 percent, would be expected to fall not more than one cup's length from the middle. You would expect about 238 (93 percent) to fall not more than two cups' length, and about 254 (99 percent) to fall not more than three cups' length, from the middle. Thus a single pinball dropped at random would have a 71 percent chance of dropping no more than one cup from the middle, and so on.

You can put this same information in a different way by predicting that a random pinball will end up no farther than one cup from the middle—if you hedge your prediction by saying you are "71 percent con-

---

* A number in any row of such a "Pascal's triangle" is the sum of the two numbers on either side just above it. (More formally, the numbers are the coefficients in the binomial expansion of $(a + b)$. Thus

$$(a + b)^4 = a^4 + 4a^3b + 6a^2b^2 + 4ab^3 + b^4$$

fident" of its correctness. If you want more confidence than this, you must settle for less precision.

The Normal Curve

As you add more and more rows and cups, you can still make such predictions—but with spread measured in "standard deviations" rather than cups. This is a measure of the variability in a set of data, usually denoted by the Greek letter sigma (σ). If you buy 1,000 packages of a product averaging 40 ounces and are told the packages vary in weight from 36 to 44 ounces, you don't know whether most of them are several ounces off the average or just a few have these extreme weights. You could express how far the average package was from 40 ounces by measuring the difference of each package from 40 ounces and averaging these 1,000 differences. Standard deviation is almost this—with the refinement that you square each difference, average the 1,000 squared values, and take the square root of the average.*

As the number of cups approaches infinity, the spread of pinballs approaches a mathematical curve known as the normal probability distribution, shown below the 9-cup case in Fig. 2-3. As the figure shows, 68 percent of all the pinballs are no more than 1 standard deviation from

* The set (8, 9, 10, 11, 12) and the set (1, 6, 10, 14, 19) have the same average value—10—but their respective standard deviations are 1.4 and 6.2.

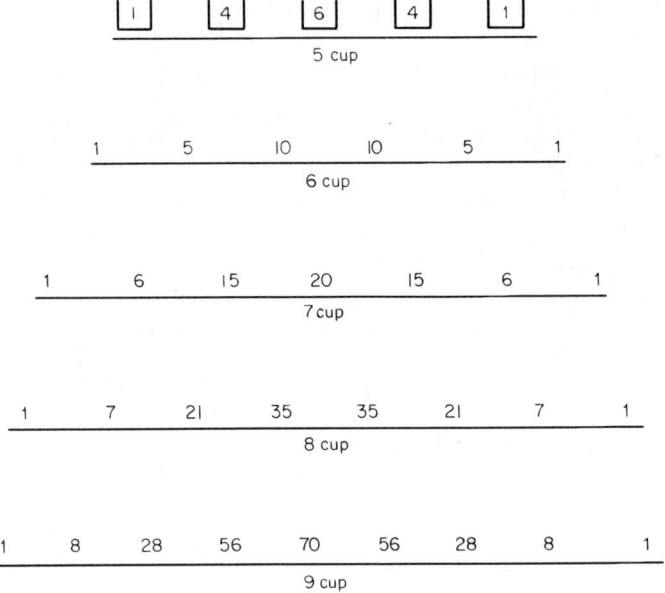

Fig. 2-2

STATISTICAL UNDERPINNINGS OF THE NEW MANAGEMENT

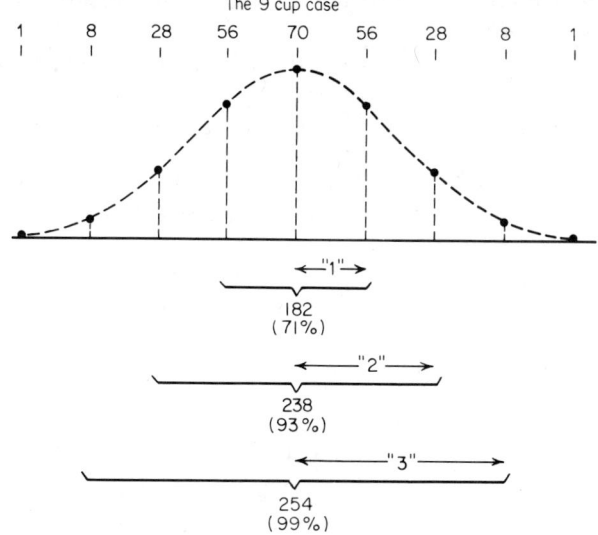

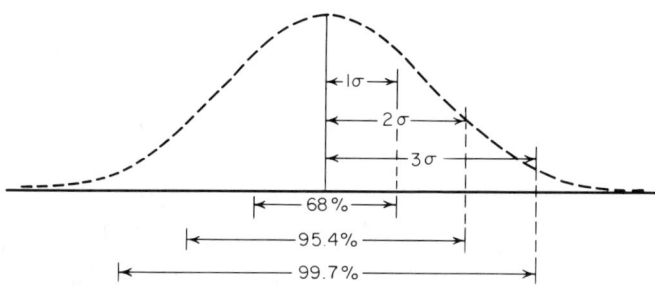

Fig. 2-3 The normal curve.

the average or middle point; and the other percentages are quite similar to the 9-cup distribution.

When populations or processes are distributed approximately this way, you can draw useful conclusions about the distance of a random value from the average. If you know the standard deviation of some large set of values distributed in this way, and if you select a single value at random:*

---

* Table 2-b shown on p. 37 tabulates for the normal distribution the fraction of all values lying between the central ordinate and any distance out on one side (expressed in standard deviations). A standard deviation of 1.96, read in the left margin of the table, corresponds to a fraction of 0.47500 of values between the central ordinate and a distance out of 1.96 (nearly 2) standard deviations; thus twice this fraction, or 95 percent, would lie approximately between 2 standard deviations on either side.

Two-thirds of the time it is within 1 standard deviation of the average value of the whole set.

Ninety-five percent of the time it is within 2 standard deviations of the average.

Almost certainly it is within 3 standard deviations of the average.

This single random selection gives you an estimate of the overall average value to some specified precision, and within some specified degree of confidence.

This may not be good enough. If you need an estimate that is within 1 standard deviation (assuming you know the units of this) and at the same time is to a confidence higher than "two-thirds of the time"—perhaps you need to be 95 percent sure—you must go to the methods of the next section.

How averages behave

The Effect of Sample Size

Suppose you want to estimate at minimum cost the average age of a group. Their ages (unknown to you) are 20, 21, 23, 24, and 27—average 23, corresponding to individuals A, B, C, D, and E.*

Suppose you make your estimate by picking a random sample of one person; how good is it?

Three of five possible samples (60 percent) are within 2 years of the correct average age.

Four of five possible samples (80 percent) are within 3 years.

All possible samples (100 percent) are within 4 years.

If this is not good enough, choose two ages at random and use the average age of this sample as your estimate. The 10 possible *samples of two* are listed in Table 2-1.

Six of ten possible samples (60 percent) have an average within 1 year of the true population average value.

Eight of ten possible samples (80 percent) have an average within 2 years.

All possible samples (100 percent) have an average within 2½ years.

This is much better: your range of error for 60 percent confidence is halved, and for higher confidences is improved markedly. Figure 2-4

---

* This population is purposely made very small, to show the concept more clearly. Imagine that it is a very large population, so that tabulating its ages would be far too costly.

STATISTICAL UNDERPINNINGS OF THE NEW MANAGEMENT 21

Table 2-1

Sample selected	Sample average
20, 21	20.5
20, 23	21.5
20, 24	22
20, 27	23.5
21, 23	22
21, 24	22.5
21, 27	24
23, 24	23.5
23, 27	25
24, 27	25.5

shows the continued improvement you would gain from further increases in sample size.

These samples are artificially small. If, to estimate the average (or other value) of a large population to a given degree of confidence, you take a random sample of reasonable size—say 20—a sample size 4 times as big will halve the range of error, and a sample size 9 times as big will reduce the range of error to a third.*

Sample versus Census

This is a good time to knock down the idea that small samples really aren't much good, and that only a complete census will do. As a practical matter, the reverse may be true. If you need an opinion survey on the desirability of a proposed bond issue in a city of 100,000 voters you could sample a random cross section of 400 voters or you could poll the entire 100,000. The first method, if properly done, would give you a result that was accurate within about 4 percent† to a fairly high confidence; the second method, *if properly done*, would be completely accurate. But that's the rub. With a small sample you can use trained professionals and precise techniques. When you survey a large population, you must use large numbers of ill-trained interviewers, who often ask biased questions, make mistakes, and falsify their reports. A huge post-war survey of industry in France, attempting to cover every business

---

* The error, in other words, is inversely proportional to the square root of the sample size.

† If your sample estimate was "36 percent in favor," say, then ±4 percent would be quite acceptable. If it came out "48 percent in favor," this would tell you a much larger sample was required, to provide the increased discrimination needed to predict the outcome of such a close vote.

n = 2			n = 3		n = 4	
A, B	$\dfrac{(20 + 21)}{2}$	20.5	A, B, C	$21\tfrac{1}{3}$	A, B, C, D	22
A, C		21.5	A, B, D	$21\tfrac{2}{3}$	A, B, C, E	$22\tfrac{3}{4}$
A, D		22.0	A, B, E	$22\tfrac{2}{3}$		
A, E		23.5	A, C, D	$22\tfrac{1}{3}$	A, B, D, E	23
B, C		22.0	A, C, E	$23\tfrac{1}{3}$		
B, D		22.5	A, D, E	$23\tfrac{2}{3}$	A, C, D, E	$23\tfrac{1}{2}$
B, E		24.0	B, C, D	$22\tfrac{2}{3}$		
C, D		23.5	B, C, E	$23\tfrac{2}{3}$	B, C, D, E	$23\tfrac{3}{4}$
C, E		25.0	B, D, E	24		
D, E		25.5	C, D, E	$24\tfrac{2}{3}$		
		230.0		230.0		115.0
Sample means		23		23		23

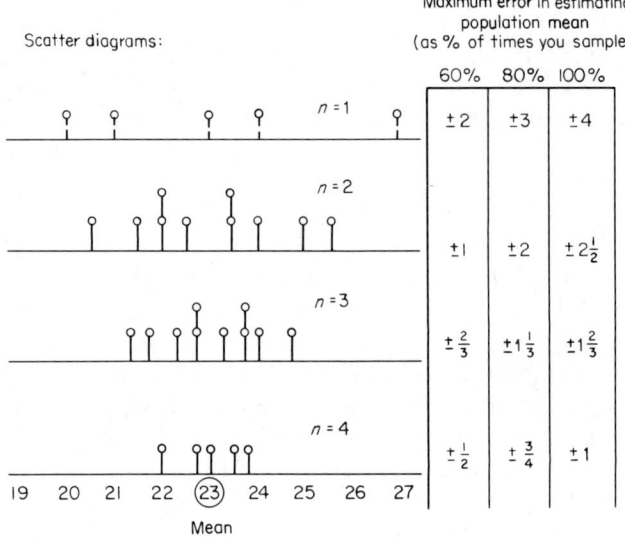

Fig. 2-4 Scatter diagrams.

above a certain size, had a final error many times what a carefully designed sample would have given.

Managers know that a census costs far more than a sampling, but they think this extra money buys them extra accuracy. Often they get less accuracy rather than more; and even when they get more, it may not be worth the cost. Discount department stores or auto parts stores which take monthly accounting inventories could accept a 5 percent error, which sampling can achieve easily. Instead they take a full inven-

Table 2-2

Size of random sample	Maximum expected error of sample percentage as an estimator of population percentage %
25	16
100	8
400	4
900	2¾
1,600	2
2,500	1⅝
6,400	1

tory which costs far more, interferes more with operations, and probably is no more accurate anyway.

If you take a random sample from a two-part population to estimate the overall proportion in each part (what proportion prefers one of two candidates, or has an income above a specified amount, or shops at a certain store), and want your estimate to have 90 percent confidence, Table 2-2 indicates approximately how large a sample you need so that the sample percentage will be within the indicated range of the actual population percentage. If 42 percent of a random sample of 1,600 from a much larger population prefer product A to product B, for example, you can be 90 percent sure that the population percentage preferring product A is within 2 percentage points of 42 percent—between 40 percent and 44 percent.

If samples are this good, why is our Census a census? Because sample size required for a given accuracy goes up swiftly as the number of questions increases. If you ask 2 questions with 3 possible answers each (such as asking whether one is old, middle-aged, or young, and whether one uses much, little, or none of a product), there are 9 answer categories. If you ask 6 questions with 5 possible answers each there are over 15,000 categories—and a sampling of 10,000 may be far too small.

Distribution of Averages

When you use a sample average to estimate the average value of the population from which it was drawn, you see from Fig. 2-4 that the precision of your estimate depends on sample size. It depends also on the *variability of the population*, and the more variable the population is, the more you have to compensate by increasing sample size. (If you want to know to some given precision where a rifle actually fires when the rifleman aligns the sights on the bull's-eye, the better shot he is, the fewer sighting shots are needed.) If the ages of your group in Fig. 2-4 had varied between 13 and 33 instead of between 20 and 27, the range of error for samples of various sizes would have been 2 or 3 times greater.

If you select a random sample of 25 and use the mean of the 25 values as your estimator, the result is only one of perhaps millions of such samples of 25 whose averages disperse themselves in a bell-shaped "normal" curve about the population average—a "distribution of sample means." This distribution has a *standard deviation*, just as does that of Fig. 2-3; 68 percent of all samples you might have picked have a mean within 1 standard deviation of the right answer, 95 percent within 2, and so on. If you know the numerical value of this standard deviation, you know at once the precision of your estimate:

> You are 68 percent confident the population average is within 1 standard deviation of your sample mean.
> You are 95 percent confident the population average is within 2 standard deviations, and so on.

If you knew the population standard deviation (σ) and the sample size (n), you could calculate this directly as follows:

> Standard deviation (of the distribution of sample means) $= \dfrac{\sigma}{\sqrt{n}}$

You don't know the population standard deviation, but you can estimate it quite closely by calculating the standard deviation of the sample you have drawn (just as, after a rifleman fires 25 shots at a target, you have a very good idea of the variability of his shooting in general).

The errors in this technique are about 5 percent for sample sizes of 20 or so, and they decrease as sample size increases.

PITFALLS IN SURVEYING

Design of a survey

It may seem the simplest thing in the world to undertake a survey. List the questions you want answered, go out and ask a good cross section of people, and the job is done. Unfortunately, as many businessmen have learned, this direct approach usually doesn't work.

Despite your best efforts, an initial set of questions on a new topic will contain bugs. If you ask "What would you like to see changed in my store?", an answer of "Nothing" may mean a respondent likes everything or just doesn't care. If you ask "How many rooms in your house are carpeted?", some respondents may think you mean wall-to-wall carpeting while others think you mean rugs. If you ask "What meat did you serve for dinner last night?", answers will be very untypical if it's just before payday. If you ask "How much do you pay each month in interest?", most respondents just don't know. If you ask "What school does your oldest child attend?", this may mean the oldest still in school or simply the oldest. You need a "pilot study," which is a small-scale replica of what you think the full survey will be, to clear up confusing questions and deal with unexpected responses.

After you're sure the answers are responsive, are you sure you want them? Your survey has an aim, but will the answers achieve it? Surveys may clear up all sorts of minor points, but neglect a principal area—as a filling station attendant may check tires, battery, windows, and radiator, but forget gasoline. The only safeguard is to carry through a complete analysis with your pilot study, and see what you overlooked.

A major contribution of the pilot study is telling you how big a sample you need. For a survey seeking a yes-no response ("Are you a Democrat? Are you a Republican?"), Table 2-2 shows the sample size needed for a given precision and 90 percent confidence, but if you want the size of something ("What do you spend per week on clothing?"), the precision of your estimate depends on the inherent variability in responses as well as on sample size, as discussed in the preceding section. The pilot study, by getting an estimate of population standard deviation, enables you to determine how large n must be to produce the desired precision.

Questionnaire errors

Questionnaires offer rich opportunities for error. Questions can be vague (what is included in "income"?), leading or slanted (does the respondent favor a "good government amendment"?), unpriced (respondents who favor colored telephones may not be willing to pay more for them), too long or complicated, an invasion of privacy, a blow at pride, and so forth. They can present alternatives in an order that favors one over others. They can give too few or too many choices.* They can indicate approval of one response over another.

---

* An example of too many choices: You ask for a first and second reason for residence moves, and give these choices: "Old house not adequate," "New house bigger," "To improve neighborhood," "To get closer to work." People who check the first two as first and second reasons are giving you the first reason twice, and not telling you their second reason at all.

A good test of a questionnaire is to present it to a group whose characteristics are known, and see if it confirms the known conditions.

Response errors

Since the rich and the poor respond less than the rest of the population, surveys undercount these groups. Mobile population of young or chronically unemployed are hard to find. Women hesitate to tell certain things to male interviewers. Certain population groups seldom answer the door (husband and wife both working, the sick, older people, women living alone). The self-employed tend to be suspicious of questions and reticent in giving business information. Professionals respond well because they appreciate the aims of surveys; housewives because adult conversation is a welcome break in their routine. These differences in response rate bias survey findings markedly unless allowances are made for them.

Interviewer errors

The most important person to your survey is the interviewer. His persistence or patience counteracts shortcomings discussed above; his dishonesty or carelessness produces fictitious data. He may have a bias against "poor homes" or upstairs residences. He may interpret the meaning of questions in a wrong way, show approval of a certain line of answers, or ask leading questions. He may ask questions of different types of people in different ways, unintentionally inviting different answers. If he is doing "quota sampling" (specified proportions of different types of respondents), he may succumb to the temptation of substituting one type for another to fill his quota.

If the interviewer is a professional who understands what you are trying to do, and if he is not pushed to get more production than is reasonable under the circumstances, he will do a good job. If he is a cut-rate employee, he can do your survey real harm.

PITFALLS IN INTERPRETING

What do the numbers mean?

Managers can spend large sums collecting and tabulating numerical information about operations, but define the meaning of the numbers quite loosely. The procurement department meets "95 percent of all requests for material," but the denominator of this ratio is undefined;

is this 95 percent of all material anyone wanted, 95 percent of all anyone asked for, 95 percent of all for which a formal requisition was submitted to the department, or 95 percent of all for which an order left the plant? The transportation department had a 75 percent equipment utilization rate; does this mean that the average vehicle was moving 18 hours a day, or that 75 percent of the equipment was checked out to a driver, or that 75 percent of the equipment was not in the repair shop? The rejection rate of a department is only 3 percent; does this mean 97 percent of production passed all inspection the first time, 97 percent passed after reworking, or 97 percent of the production that got by shop inspection passed final inspection?

When data seem to be saying one thing but actually say something else, management tries to correct a situation which does not exist. The President's Commission on Law Enforcement and the Administration of Justice* noted that the FBI crime index made no distinction with respect to seriousness; thus it showed crime in the District of Columbia on the increase because the total number of homicides and car thefts combined went up, whereas a reasonable index of seriousness would have reported a drop in the weighted index. It is worthwhile to analyze performance measures with great care before they creep into management reports.

What is being measured?

The process of measurement often changes what is being measured. Engineers encounter this in a small way: the device that records vehicle speed requires a tiny bit of power which slows the vehicle a bit; the strength test stresses the part and drains off a bit of strength. In management the changes can be substantial. When the boss measures certain elements as an index of performance, he has to recognize the strong tendency for those elements to be maximized at the expense of others he does not (perhaps cannot) measure. There is a natural desire to find out what the boss wants (and what better way to do this than to note what he is measuring?), and then devote highest priority to giving it to him.

If your primary measure is profit, you can pressure subordinate managers to neglect maintenance, or training, or investment in anything without immediate return. They may gamble on reductions in supervision, safety, or quality control, creating short-term savings but leading to a gradual downhill trend or inviting expensive mishaps.

* *Task Force Report: Science and Technology*, a Report to the President's Commission on Law Enforcement and the Administration of Justice, prepared by the Institute for Defense Analysis, Government Printing Office, Washington, D.C., 1967.

If you measure overhead ratio, you invite shift of overhead costs to direct labor. If you emphasize percentage of markup achieved, you encourage large purchases to take advantage of quantity discounts, even though increased inventory costs may eat up profit. If you measure any routinized operations, you may take managers' eyes off the innovative aspects of their jobs, or the exploratory work of developing new customers. The depersonalizing effect of quantitative performance measures can do far more harm than good if measures are selected hastily or used as a substitute for personal supervision.

This is not an argument against using numbers in management analysis; indeed, this whole book is about doing just that. It is a reminder that when the activity being measured senses what you measure, there is a conscious change tending to invalidate the measure. This risk is not present when you measure inanimate operations. It may not be present if the element you are evaluating is not the major goal of an activity.* And it is not, of course, present if you are interested in absolutely nothing except the things you are measuring.

This last situation means you are evaluating a very simple operation, or you have a very effective management information system. More and more companies are moving in the latter direction, and although successes in this area are limited by mistrust and skepticism, clear progress is being made. When a company makes the decision to construct a "total" information system for management, with all primary data collected and stored in the most economical way, and with data access so versatile that almost any conceivable performance ratio can be extracted by management, defensive reaction becomes extremely difficult. In the first place, with executive analysis so unscheduled and unsystematized, it is not possible to know what performance measures will be reviewed; in the second place, the very wholesale nature of the commitment to information gathering produces a feeling of inevitability—and the human animal adjusts to this new environment of total disclosure.

There can be unwholesome social implications to a management information system, unless its installation is achieved in an atmosphere of cooperation and mutual confidence. The system should help people, not hurt them—and they should be made to know this. It is not the place of this book to explore social problems of privacy invasion, but only to note that certain social needs of an organization must be met if your management measurement is to be effective.

* An investigation of police implication in a racial killing a few years ago gave limited credence to testimony from the policemen concerned; but telephone company records of pertinent calls were taken at face value, since the company's involvement was slight in comparison with its compelling need for accurate records.

HOW TO INCORPORATE JUDGMENT

The "numbers-or-judgment" myth

When Robert McNamara first went to the Pentagon, there was a sharp clash between two irreconcilable camps: the experienced generals who had come of age amid the stench and shock of combat, and the youthful intellectuals who set out to mastermind future wars with computers and the new math. The former, who hadn't studied quantitative economics and did not propose to start now, were profoundly suspicious of the pipe-smoking mathematicians who had never been there but knew just how it should be. The latter spoke contemptuously of the nontransferability of experience to the new environment and of the naïveté of military presentations compared with quantitative analyses. The dialogue reflected little credit on either side. As time passed there was mellowing and mutual adjustment. The recalcitrant generals passed from the scene, to be replaced with a newer breed that had studied a bit of the new math. The young analysts picked up both age and humility. "In our youth we looked more scientific," said one; "now we are coming to realize that modern techniques are not powerful enough to penetrate a future seen so dimly." And another noted that "almost any analysis by a competent operations research group can be torn to shreds by an opposing group using different, but equally supportable, assumptions."

Business sometimes encounters this same split personality. On one hand may be the systems analyst, who feels that essential factors of a problem can be reduced to a mathematical model, after which the data are fed in and the best decision emerges. On the other may be the experienced executive, who has seen enough flawed solutions by staff numerologists and who has lived so long by his business judgment that he feels justified in preferring the latter over the former. This is unfortunate, for there is no reason not to use both.

The reliability of information

A happy marriage of data and judgment is hampered by the mechanical difficulty of giving proper weight to each. When a new product needs a volume of 60 a day to succeed, and survey results predict a volume of 50 but the sales manager predicts 80, some executives incline toward the apparent precision of the former while others favor the wide-gauge consideration implied by the latter. A selection of one or the other is wasteful, however, for each contains information, and information is too scarce a managerial commodity to squander.

If you combine two or more sources of information, intuition tells you the weighting should be based on the relative quality of each. The pinball trial in an earlier section showed that, when information is quantitative, in the form of a sample average used to estimate a population average, the standard deviation provides a measure of the estimate's quality. (The tight 10-shot group of an expert marksman is a more reliable estimator of the gun's point of impact than 10 widely scattered shots of a tyro.) When you collect numerical data, the method given earlier enables you to calculate the standard deviation, but you need a way to put similar reliability measures on human judgments, and a way to use these measures in weighting estimates.

The Standard Deviation of Judgment

If something is known with certainty, it is not judgment but fact. Executive judgment implies a degree of uncertainty, and you seek to express this uncertainty in numerical terms. If you pin down the sales manager by asking if he is virtually positive that daily volume will average exactly 80 units, he will say no. His best estimate is 80 units, but in his mind there is a range and a confidence. If he could say, "I am 68 percent confident that the average volume will turn out to lie between 70 and 90 units a day," he would tell you that the standard deviation of his estimate is 10 units.* This is expecting a bit too much, but it is reasonable for him to establish a range with a 50 percent confidence limit. He starts with such a wide range that he is virtually certain the average will fall inside it, then decreases the limit until he reaches a range where his judgment tells him it is a toss-up—he can't decide whether the average will fall within this range or not. If you consult a table of the normal distribution, you find that this range corresponds to two-thirds of a standard deviation. If his 50 percent confidence range is ±6 units a day (a range of 74 to 86, that is), you take 6 units to be two-thirds of the standard deviation; and the standard deviation of his estimate thus is 9 units a day.

* The assumption is that his estimating uncertainty behaves as does the normal probability distribution pictured on Fig. 2-3—often a pretty good assumption.

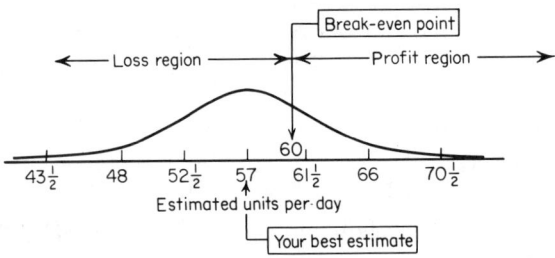

Fig. 2-5

STATISTICAL UNDERPINNINGS OF THE NEW MANAGEMENT

Combining Data and Judgment

Suppose your market survey, which puts average daily volume at 50 units, has a standard deviation of 5 units; and from the previous paragraph you assess a standard deviation of 9 units for the sales manager's estimate of 80 units a day. How do you combine these?

Conceptually, you weight each by its reliability. Arithmetically (without going through the rationale), you weight each by the reciprocal of its standard deviation squared, as follows:

$$\text{Combined volume estimate} = \frac{50 \times 1/25 + 80 \times 1/81}{1/25 + 1/81}$$

$$= 57.2 \text{ units per day}$$

The standard deviation of this combined estimate must be smaller than either of the individual standard deviations—5 or 9—if it is a more reliable estimate than either. It can be approximated by the formula for the "harmonic mean," as follows:

$$SD_{combination} = \frac{1}{\sqrt{\frac{1}{(5)^2} + \frac{1}{(9)^2}}} = 4.4 \text{ units per day}$$

Your best assessment, then, is an average daily volume of about 57 units, with a standard deviation of about 4½ units a day. Figure 2-5 plots this in the form of a normal curve.

You can estimate a 68 percent chance that your average daily volume will be between 52½ and 61½—with a 34 percent chance of its falling on either side. There is an 84 percent chance it will be 61½ or less (50 percent + 34 percent). The table of the normal probability distribution tells you that there is about 1 chance in 4 that you will attain an average daily volume above 60 units. If 60 is an accurate break-even estimate you must not enter this market. It may be time for your break-even estimate to be explored more precisely. In any case, you have made the best possible use of available information, not choosing facts or judgment but combining both.

INSPECTING, TESTING, AND QUALITY CONTROL

How much inspection?

If you are manufacturing components for a spaceship, and the cost of a defective unit is astronomical (in more ways than one), your inspection strategy is simple—inspect everything. If you are filling birdseed boxes, and simply want to keep on the right side of the weights and measures authorities, your problem is tougher. Your filling machinery has some inherent variation, and the weight of birdseed in the boxes therefore has a standard deviation; and since it costs either to inspect or to overfill enough to make inspection unnecessary, you have a problem of finding the minimum-cost strategy.

Suppose the law requires that, for a case of 16 boxes taken at random, the average weight may not fall below 8 ounces. Suppose your data tells you that the standard deviation of weight is 1 ounce. Unless you are willing to inspect every box, you must follow an overfilling policy to avoid running afoul of the law. Figure 2-6 shows this situation. Since the standard deviation for individual package weights is 1 ounce, the standard deviation used in Fig. 2-6 is for average weights of sets of 16 packages, and thus is $1/\sqrt{16}$ or $\frac{1}{4}$ ounce. The illustration shows that, to meet the requirements virtually every time an inspection is conducted, the average weight to which you must fill is $8\frac{3}{4}$ ounces. If you are willing to have a law violation about $2\frac{1}{2}$ percent of the time, you have to fill to an average of $8\frac{1}{2}$ ounces; and if you will accept 16 percent violations, you fill to an average of $8\frac{1}{4}$ ounces.

Suppose you adopt the strategy of no violations, and select an average filling weight of $8\frac{3}{4}$ ounces. Your filling machinery has a built-in random variation which you cannot avoid; but also it may go gradually out of adjustment so that the *average* weight changes. You must set up an inspection system to detect this maladjustment, so that a mechanic can readjust to the proper average weight.

Assume that the average weight changes in $\frac{1}{4}$-ounce increments, so that if it moves from $8\frac{3}{4}$ it goes first to $8\frac{1}{2}$ or 9; and that the maladjustment can occur only when starting in the morning, so that the adjustment for the first run continues throughout the day. If you inspect 64 packages every morning, what chance do you have of catching a $\frac{1}{4}$-ounce change in average weight? The standard deviation of weight for averages of 64 packages is $1/\sqrt{64}$ or $\frac{1}{8}$. Figure 2-7 shows what the distribution of average weights for 64-package lots would look like if the filling machinery were in adjustment or if it were adjusted $\frac{1}{4}$ ounce high or low. The inspection rule might be to set upper and lower control limits of $8\frac{1}{2}$ and 9,

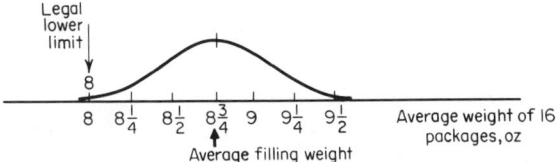

Fig. 2-6

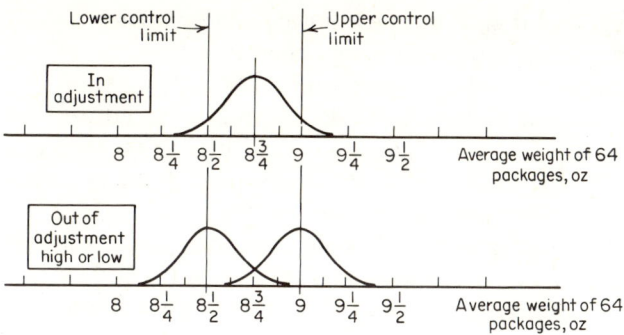

Fig. 2-7

and to readjust if the average weight of the 64 packages were outside these limits.

How selective would this inspection rule be? If the machinery actually was in adjustment, there would be a 5 percent chance of getting an average outside the control limits (leading you to make a needless adjustment). If the machinery actually was ¼ ounce out of adjustment, either high or low, there would be a 50 percent chance of getting an average inside the control limits (leading you to accept an incorrect adjustment). The situation where the adjustment was ½ ounce high or low is not shown on the illustration, but if you plotted it you would see that there would be only a 2½ percent chance of getting an average inside the control limits (thus accepting an incorrect adjustment).

This inspection rule is quite selective for differentiating between machinery in adjustment and machinery ½ ounce of adjustment; only 5 percent of the time would you adjust needlessly, and only 2½ percent of the time would you fail to correct a maladjustment. If you want a rule this selective for a difference of only ¼ ounce in average values, you must quadruple the size inspection sample you take and bring the control limits halfway in to 8¾ on each side.

Sequential sampling

The conclusions from the preceding paragraph can be stated as follows:
1. If you wish to detect a ½-ounce maladjustment with only a 2½ percent error rate (and needlessly adjust a correct setting 5 percent of the time), take a sample of 64 and set the control limits for the sample average at 8½ and 9.
2. If you wish the same discrimination for a ¼-ounce maladjustment, take a sample of 256 and set the control limits at 8⅝ and 8⅞.

There is a cheaper solution for the $\frac{1}{4}$-ounce maladjustment case; it involves a two-stage sampling procedure, as follows:
1. Take a sample of 64 and set the control limits at $8\frac{1}{2}$ and 9.
2. If the sample average falls outside the control limits, adjust.
3. If the sample average falls inside the control limits take a second sample of 192; combine this with the first sample to make a total sample of 256 and set the control limits at $8\frac{5}{8}$ and $8\frac{7}{8}$.

This sequential sampling method has the advantage that you will take many fewer samples in the long run; and the likelihood of making a wrong decision is only slightly increased. There can be far more complicated techniques than this, involving several stages; the idea behind each such technique is to take as few samples as possible and to stop sampling as soon as enough evidence is in to provide a decision to the confidence you need.

To sample or not to sample

The Problem: Decide Now or Seek More Information?

The manager often has the uncomfortable feeling that he isn't ready to make a rational choice, but he isn't sure whether the gain from more information will exceed its cost. The cost of getting information by sampling is the direct expenditure for sampling, measuring, or interviewing. The benefit is the difference between the cost he would incur from taking the optimum action without sampling and that from taking the optimum action dictated by the sample findings. The joker is that he does not know what the optimum action will be until he knows the sample findings —and he can't have them until he pays for them.

The latter is true, for a specific case, but it may not be true in the long run, and this is what makes an assessment of the benefits of sampling possible.

Rationale of the Method

You are wondering whether to sample or not, and you must make your decision before you know what the particular sample will disclose. You evaluate the expected cost you would incur after you:
1. Take a sample of some specified size.
2. Take the action indicated by the sample you probably would get. You decide the expected value of the sample you probably would get by computing:
   a. The cost of the action you would take for each possible sample outcome, times
   b. The probability of occurrence of that sample outcome

STATISTICAL UNDERPINNINGS OF THE NEW MANAGEMENT

Table 2-3

Value	Description	Probability, %
$8\tfrac{1}{4}$	$\tfrac{1}{2}$ oz low	10
$8\tfrac{1}{2}$	$\tfrac{1}{4}$ oz low	20
$8\tfrac{3}{4}$	Adjusted	40
9	$\tfrac{1}{4}$ oz high	20
$9\tfrac{1}{4}$	$\tfrac{1}{2}$ oz high	10

Example of the Method

Go back to the birdseed example. Suppose there are five possible adjustments of the filling machinery; these, with the probability of occurrence of each (obtainable from past observations) are given in Table 2-3. Figure 2-6 and its associated text tell you that the penalty for an $8\tfrac{1}{2}$-ounce average weight is a law violation $2\tfrac{1}{2}$ percent of the time, and the penalty for an $8\tfrac{1}{4}$-ounce average weight is a law violation 16 percent of the time. The penalty for a 9-ounce average weight is a wastage of $\tfrac{1}{4}$ ounce of birdseed per box all day, and the penalty for a $9\tfrac{1}{4}$-ounce average is the wastage of $\tfrac{1}{2}$ ounce of birdseed per box all day. Suppose you were able to put prices on these departures from the correct weight, as shown in Table 2-4. The expected daily cost of weight error, got by multiplying each cost by its probability of being incurred and summing, is $4.90. Suppose the daily cost of adjusting the machinery, whether needed or not, is $4.20. Since you expect to lose, on the average, $4.90 a day if you do not adjust, the rational action in the absence of sampling is to adjust every day, which saves you 70 cents a day over your cost if you do not sample.

If you sample, what will the act of sampling cost you? You will take a sample of 64 every day, and 59 percent of the time you will take an additional sample of 192 (this calculation, not shown, is somewhat similar to that of Table 2-5); on the average, therefore, you will take a sample of

Table 2-4

Average weight	Probability	Daily conditional cost of weight error	Daily expected cost of weight error
$8\tfrac{1}{4}$	0.1	$16.00	$16.00 × 0.1 = $1.60
$8\tfrac{1}{2}$	0.2	2.50	$2.50 × 0.2 = $0.50
$8\tfrac{3}{4}$	0.4	.00	
9	0.2	4.00	$4.00 × 0.2 = $0.80
$9\tfrac{1}{4}$	0.1	20.00	$20.00 × 0.1 = $2.00
			$4.90

Table 2-5

Average weight	Probability	Better act	Conditional cost of better act	Expected cost of better act (cost $\times$ prob^y)
$8\frac{1}{4}$	0.1	Adjust	$4.20	$0.42
$8\frac{1}{2}$	0.2	Don't adjust	2.50	0.50
$8\frac{3}{4}$	0.4	Don't adjust	.00	.00
9	0.2	Don't adjust	4.00	0.80
$9\frac{1}{4}$	0.1	Adjust	4.20	0.42
			Nonsampling cost of sampling	$2.14
			Sampling cost of sampling	1.76
			Total expected cost of sampling	$3.90 per day

176 boxes. If sampling costs 1 cent a box, the daily sampling cost will be $1.76.

If you sample, what will your other costs be every day? (Ignore the fact that your sampling is not 100 percent accurate, to simplify the problem; in a real case you would take this factor into account.)

When the average weight is estimated to be $8\frac{1}{2}$, $8\frac{3}{4}$, or 9, you will not adjust, since the average daily cost of these weight errors is less than the cost of adjustment. When the average weight is estimated to be $8\frac{1}{4}$ or $9\frac{1}{4}$, you will save money if you adjust. The expected cost of sampling is computed in Table 2-5.

In the long run, if you sample every day and take the optimum action indicated by the sample, your cost will be $0.30 per day ($4.20 − $3.90) less than if you take the optimum action without sampling—that is, if you adjust every day.

This technique, which in practice may be more elaborate but is no different in concept, is known as "preposterior analysis." It holds obvious advantages whenever your process, although random regarding individual outcomes, follows some distribution pattern which you can approximate.

EVOLUTIONARY OPERATIONS

It is one thing to get as much data as possible about where your business has been in the past. It is entirely different to find out where it is going in the future. If your past operations have been confined to doing exactly what you thought was optimum all the time, you have a vast pile of statistics about one narrow mode of operation, and no information at all about the effects of different strategies. Mathematical models and statistical analysis can provide considerable information, but not in a vacuum.

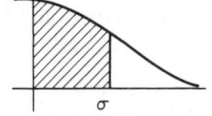

Table 2-6 Areas of the normal curve between the maximum ordinate and successive values of standard deviation (σ)*

σ	00	01	02	03	04	05	06	07	08	09
0.0	00000	00399	00798	01197	01595	01994	02392	02790	03188	03586
0.1	03983	04380	04776	05172	05567	05962	06356	06749	07142	07535
0.2	07926	08317	08706	09095	09483	09871	10257	10642	11026	11409
0.3	11791	12172	12552	12930	13307	13683	14058	14431	14803	15173
0.4	15542	15910	16276	16640	17003	17364	17724	18082	18439	18793
0.5	19146	19497	19847	20194	20540	20884	21226	21566	21904	22240
0.6	22575	22907	23237	23565	23891	24215	24537	24857	25175	25490
0.7	25804	26115	26424	26730	27035	27337	27637	27935	28230	28524
0.8	28814	29103	29389	29673	29955	30234	30511	30785	31057	31327
0.9	31594	31859	32121	32381	32639	32894	33147	33398	33646	33891
1.0	34134	34375	34614	34850	35083	35314	35543	35769	35993	36214
1.1	36433	36650	36864	37076	37286	37493	37698	37900	38100	38298
1.2	38493	38686	38877	39065	39251	39435	39617	39796	39973	40147
1.3	40320	40490	40658	40824	40988	41149	41309	41466	41621	41774
1.4	41924	42073	42220	42364	42507	42647	42786	42922	43056	43189
1.5	43319	43448	43574	43699	43822	43943	44062	44179	44295	44408
1.6	44520	44630	44738	44845	44950	45053	45154	45254	45352	45449
1.7	45543	45637	45728	45818	45907	45994	46080	46164	46246	46327
1.8	46407	46485	46562	46638	46712	46784	46856	46926	46995	47062
1.9	47128	47193	47257	47320	47381	47441	47500	47558	47615	47670
2.0	47725	47778	47831	47882	47932	47982	48030	48077	48124	48169
2.1	48214	48257	48300	48341	48382	48422	48461	48500	48537	48574
2.2	48610	48645	48679	48713	48745	48778	48809	48840	48870	48899
2.3	48928	48956	48983	49010	49036	49061	49086	49111	49134	49158
2.4	49180	49202	49224	49245	49266	49286	49305	49324	49343	49361
2.5	49377	49396	49413	49430	49446	49461	49477	49492	49506	49520
2.6	49534	49547	49560	49573	49585	49598	49609	49621	49632	49643
2.7	49653	49664	49674	49683	49693	49702	49711	49720	49728	49736
2.8	49744	49752	49760	49767	49774	49781	49788	49795	49801	49807
2.9	49813	49819	49825	49831	49836	49841	49846	49851	49856	49861
3.0	49865	49869	49874	49878	49882	49886	49889	49893	49897	49900
3.1	49903	49906	49910	49913	49916	49918	49921	49924	49926	49929
3.2	49931	49934	49936	49938	49940	49942	49944	49946	49948	49950
3.3	49952	49953	49955	49957	49958	49960	49961	49962	49964	49965
3.4	49966	49968	49969	49970	49971	49972	49973	49974	49975	49976
3.5	49977	49978	49978	49979	49980	49981	49981	49982	49983	49983
3.6	49984	49985	49985	49986	49986	49987	49987	49988	49988	49989
3.7	49989	49990	49990	49990	49991	49991	49992	49992	49992	49992
3.8	49993	49993	49993	49994	49994	49994	49994	49995	49995	49995
3.9	49995	49995	49996	49996	49996	49996	49996	49996	49997	49997
4.0	49997	49997	49997	49997	49997	49997	49998	49998	49998	49998

* Read each area value as a fraction of 1.00000; thus 34134 would be read 0.34134, or approximately 34%.

Obviously it is not sensible to vary your business practices drastically, or to take risks that can lead to heavy losses, just for the sake of collecting statistics. But the manager encounters many crossroads where one path seems almost as good as another. When he reaches such a junction, it is time to think of the guidance he can get for the future if he tries more than one road to profit.

"Evolutionary operations" is the name for a style of business management that recognizes the unique value to you of past records about *your* business and seeks to make them many times more valuable by carefully designed experimentation within the broad range of good business practice. It is well for all levels of management to keep this concept generally in mind; but top management should design consciously and specifically for such experimentation. In the absence of data produced in such a way, top management is denying itself an important element of intelligence that it could have at very little cost. In certain areas it will be forced to fly blind; and blind management flying is an extravagant sport.

3
Business Strategies for Increased Profit

Profits are central to the aim of any business. Your business has been profitable overall, or it would not stay alive; but it has not been equally profitable each month. More to the point, it has not always been as profitable as it could have been.

How can you increase profits? This is your constant question, and doubtless you have experimented with many merchandising tricks—promotions, markdowns, advertising, changing lines—in your effort to find the magic formula. As your strategy varied, your profits went up or down, presumably giving you a clue to the best tactics. But it wasn't that simple. You changed more than one thing at a time, so it wasn't clear whether a good month resulted from heavy advertising or the fashion show. The payoff in profits wasn't consistent; sometimes advertising boosted sales and sometimes it didn't. Outside factors were a complication: layoffs at the local factory, the general state of the economy, variation in demand over the seasons. The sum total is a set of books with no clear cause-and-effect pattern at all. Or so it seems.

Such a pessimistic conclusion is unwarranted, however; your records contain priceless management data about your business, and the information you seek is in them. The statistical tricks for extracting it are logical and understandable, and you may as well learn them.

Consider your average net profit as a combination of three forces:

Average net profit	=	basic profit potential	±	internal strategy effect	±	external economy effect

Basic profit potential is compounded of your available market, your advertising level on the average, your merchandising skill, the long-run business climate, the general level of promotions, and so on. Year in and year out, it is the "expected profit" of a business like yours, run in a normal way.

Internal strategy effect is the increase or decrease in your basic expected profit because of conscious variations in your business strategy. If you were sure some business strategy paid off best, you would raise your budget for this strategy—and this would produce an increase over the basic expected profit.

External economy effect is a separate increase or decrease in net profit, apart from what you achieve by your own efforts, because of what the economy is doing. You can't influence this change, but if you can see it coming you can plan accordingly.

In this chapter you'll estimate the combined result of these first two effects—basic profit potential modified by internal strategy—ignoring the fluctuations in the economy (in effect, assuming average business conditions). In Chap. 4 you will include the additional effect produced by the changing economy.

SEASONAL ADJUSTMENT

Assume you have profit and cost records for six years past, as in Table 3-1. If you compare net profit for a February when you launched a marketing campaign with that for an August when it was just "business as usual," you can be misled badly. You aren't trying to compare February with August—the latter may be typically a good month and the former a slow one. You want to compare the February of your heavy campaign with the average February. If you raised February's average $2,000 profits to $4,000, the campaign paid off; but comparing this $4,000 with August's $5,000 average profit hides this solid achievement.

However, the system of comparing the same calendar months won't do—you'd have to wait several years to assess the payoff of any new strategy. Besides you're wasting valuable past data about your operations. You need to use *all* months for your comparisons, but with seasonal variations washed out. If February profits are 40 percent of August's on the average, $2 made in February is an achievement equal to

Table 3-1

Year	Month	1961–1963				Year	1964–1966			
		Net profit	Promotion costs	Advertising costs	Net profit (seasonally adjusted)		Net profit	Promotion costs	Advertising costs	Net profit (seasonally adjusted)
1961	Jan.	(−) 5,700	7,200	4,500	1,000	1964	(−) 4,500	2,500	5,100	2,800
	Feb.	4,300	3,500	8,000	7,000		1,600	900	7,100	4,000
	Mar.	1,200	4,900	5,000	1,400		8,100	2,100	6,300	4,600
	Apr.	9,900	5,200	5,800	8,500		600	3,900	7,300	400
	May	(−) 600	8,300	6,800	2,200		600	6,700	6,200	(−) 1,100
	June	800	4,400	4,900	0		2,100	3,900	5,100	1,700
	July	(−) 1,600	5,200	3,000	2,800		7,700	3,900	9,400	600
	Aug.	(−) 1,700	3,200	4,400	4,100		2,600	3,500	7,200	10,800
	Sept.	5,200	1,500	5,600	4,000		8,300	1,200	8,700	1,500
	Oct.	7,500	5,900	4,600	4,600		10,900	900	8,600	5,200
	Nov.	(−) 6,200	2,200	5,500	11,800		7,800	11,100	11,600	1,400
	Dec.	1,200	1,900	4,400	11,400		(−) 7,800	1,100	2,300	300
1962	Jan.	(−) 100	2,700	4,100	9,300	1965	(−) 6,600	6,700	7,800	300
	Feb.	6,000	1,500	6,400	8,900		400	2,100	8,800	1,800
	Mar.	3,500	2,300	4,400	600		19,000	3,400	10,000	14,200
	Apr.	5,700	2,500	5,500	4,500		5,900	1,300	6,900	3,800
	May	4,600	4,100	6,100	7,200		11,800	2,100	9,800	9,200
	June	6,100	900	4,900	2,600		6,200	7,100	9,800	7,300
	July	(−) 8,500	1,900	1,600	9,300		(−) 10,100	3,500	9,800	8,100
	Aug.	4,600	2,800	4,700	3,000		4,000	1,200	6,600	2,200
	Sept.	(−) 11,400	5,100	5,500	3,400		3,700	4,900	8,600	4,500
	Oct.	(−) 7,500	2,700	6,700	12,300		14,600	2,300	10,500	10,900
	Nov.	(−) 9,400	1,500	6,900	900		17,000	13,100	9,400	5,700
	Dec.			2,300	2,300		1,100	1,100	4,500	15,200
1963	Jan.	4,100	2,000	4,300	3,400	1966	(−) 2,800	800	8,300	5,300
	Feb.	5,800	2,400	4,400	4,200		2,900	600	12,100	5,500
	Mar.	5,100	2,400	4,900	2,500		12,500	1,900	10,500	8,500
	Apr.	3,000	4,100	5,900	3,800		12,900	3,800	9,200	11,300
	May	6,400	14,700	5,000	4,200		14,500	7,100	11,900	11,600
	June	2,600	6,400	4,600	4,500		3,800	2,100	9,700	4,800
	July	2,500	14,800	7,400	1,000		29,400	3,800	11,100	26,100
	Aug.	5,500	1,100	5,400	(−) 3,900		9,400	2,900	10,200	12,700
	Sept.	(−) 1,700	1,000	4,300	600		21,100	8,100	7,000	19,200
	Oct.	3,700	5,200	7,700	1,200		19,100	7,300	11,300	14,900
	Nov.	(−) 7,200	3,300	8,500	1,100		38,800	4,400	11,800	20,800
	Dec.	(−) 7,400	2,000	2,400	1,000		(−) 9,000	500	2,900	1,700

$5 made in August; so before you compare this February's $4,000 profit with a typical August profit, you divide by 0.4—and compare $4,000/0.4, which is $10,000, with the $5,000 for August. This is the idea behind seasonal adjustments for all 12 months. Find January's average profit over six years, February's average, and so on, as in Table 3-2. Now see how your average January compares with the all-months' average, and the ratio between the two is a seasonal adjustment factor for January.

An example will make it clear. Your average November performance has been (from Table 3-2) a profit of $14,600, and your average December performance a *loss* of $5,620. (Since you have had occasional months with losses larger than $10,000, and you want all positive values for calculating seasonal adjustment factors, add $20,000 to each monthly profit figure—but don't forget to subtract it out again when you're done.)

November seasonal adjustment factor
$$= \frac{\text{November average profit} + \$20{,}000}{\text{all-months' average profit} + \$20{,}000} = \frac{\$14{,}600 + \$20{,}000}{\$3{,}970 + \$20{,}000}$$
$$= \frac{\$34{,}600}{\$23{,}970} = 144\%$$

December seasonal adjustment factor $= \dfrac{-\$5{,}620 + \$20{,}000}{\$3{,}970 + \$20{,}000}$
$$= \frac{\$14{,}380}{\$23{,}970} = 60\%$$

Seasonal adjustment factors for all months are shown in Table 3-2.

Before you really know how well you did in any month, you must deseasonalize your profits. How well did you do in November and December of 1961? The November, 1961, net profit of $6,200 must be increased by $20,000, divided by 144 percent, then decreased by $20,000;

$$\frac{\$6{,}200 + \$20{,}000}{1.44} = \$18{,}200$$

and subtracting $20,000 leaves −$1,800.

Your *real* November 1961 performance, adjusting for the fact that you ought to do quite well in November, is *a loss of* $1,800, *relative to your usual November performance.* Your *real* December 1961 performance is: (−$1200 + $20,000)/0.60 − $20,000, or *a profit of* $11,400 *relative to your usual December performance.* You did far better in December, considering the unfriendliness of the season for your product, than you did in November—and any special marketing efforts you made should be judged on the basis of *how much more profit you made than you usually make that month.*

After you have seasonally adjusted each of your 72 monthly profit figures, as shown in the right-hand column of Table 3-1, you can now

Table 3-2 Seasonal adjustment factors

	Year	Month					
		Jan.		Feb.		Mar.	
	1961	(−)	5,700		4,300		1,200
	1962	(−)	100		6,000		3,500
	1963	(−)	4,100	(−)	5,800	(−)	100
	1964	(−)	4,500		1,600		8,100
	1965	(−)	6,600	(−)	400		19,000
	1966	(−)	2,800		2,900		12,500
Total		(−)	23,800		8,600		44,200
Average		(−)	3,970		1,430		7,370
Seasonal adjustment factor			0.68		0.90		1.14
		Apr.		May		June	
	1961		9,900	(−)	600	(−)	800
	1962		5,700		4,600		6,100
	1963	(−)	3,000		6,400		2,600
	1964		600		600		800
	1965		5,000		11,800		6,200
	1966		12,900		14,500		3,800
Total			31,100		37,300		18,700
Average			5,180		6,220		3,120
Seasonal adjustment factor			1.05		1.09		0.96
		July		Aug.		Sept.	
	1961	(−)	1,600		1,700		5,200
	1962	(−)	8,500		500		4,600
	1963		2,500	(−)	5,500		1,700
	1964		2,100		7,700		2,600
	1965		10,100	(−)	4,000	(−)	3,700
	1966		29,400		9,400		21,100
Total			34,000		9,800		31,500
Average			5,670		1,630		5,250
Seasonal adjustment factor			1.07		0.90		1.05
		Oct.		Nov.		Dec.	
	1961		7,500		6,200	(−)	1,200
	1962	(−)	11,400		7,500	(−)	9,400
	1963		3,700		7,200	(−)	7,400
	1964		8,300		10,900	(−)	7,800
	1965		14,600		17,000		1,100
	1966		19,100		38,800	(−)	9,000
Total			41,800		87,600	(−)	33,700
Average			6,970		14,600	(−)	5,620
Seasonal adjustment factor			1.12		1.44		0.60
All-months' average					3,970		

compare any month's performance directly with any other for assessing the effect of your marketing strategies.

VISUAL ANALYSIS

Now you are ready to see which actions have increased or decreased the seasonally adjusted net profit.* You consider only those actions that

* Henceforth in this discussion, when the term "net profit" is used it will be understood to mean seasonally adjusted net profit.

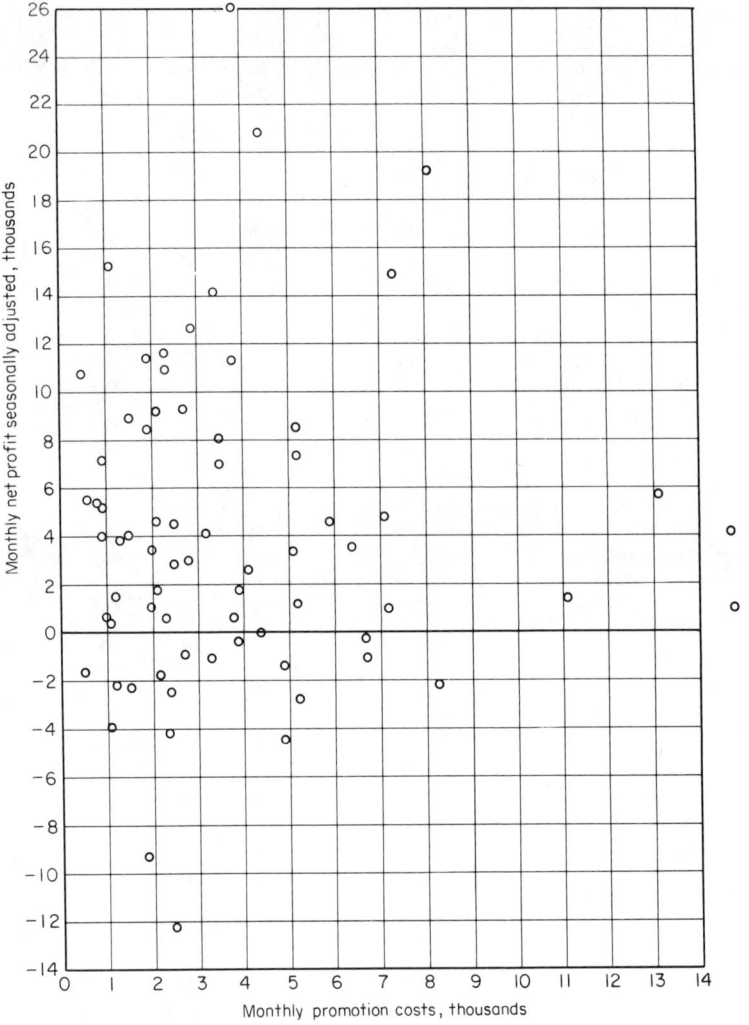

Fig. 3-1 Net profit vs. promotion cost (monthly).

BUSINESS STRATEGIES FOR INCREASED PROFIT 45

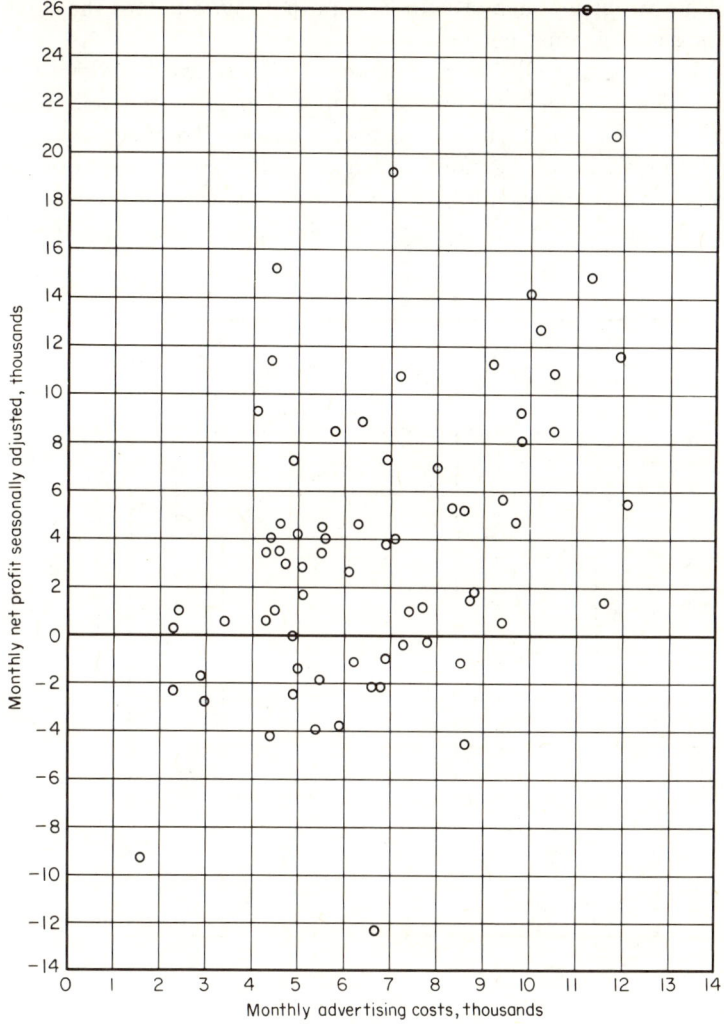

Fig. 3-2 Net profit vs. advertising cost (monthly).

you can price, passing up intangibles such as courtesy by salespeople which are important but aren't directly relatable to costs. The place to find internal strategies that carry a price is in the books; any expenditure item you've carried separately, and have varied up and down on a judgmental basis, is a candidate for investigation.

Suppose you've narrowed the choice to two: promotions and advertising. Table 3-1 contains promotional costs, advertising cost, and net profit for each month of the six years. In order to see how closely each of these relates to profit, make a plot of each. Figure 3-1 plots promotional cost against net profit for each of your 72 months; Fig. 3-2 does the same for advertising cost against net profit.

Figure 3-1 gives virtually no sign of correlation between promotions and profits. The months in which you spent the most brought no more net profit than those in which you spent the least. To check for any underlying trend, separate the months into groups by size of promotional expenditures—12 months with lowest expenditures in the first group, the next lowest 12 in the second, and so on—and plot these six average

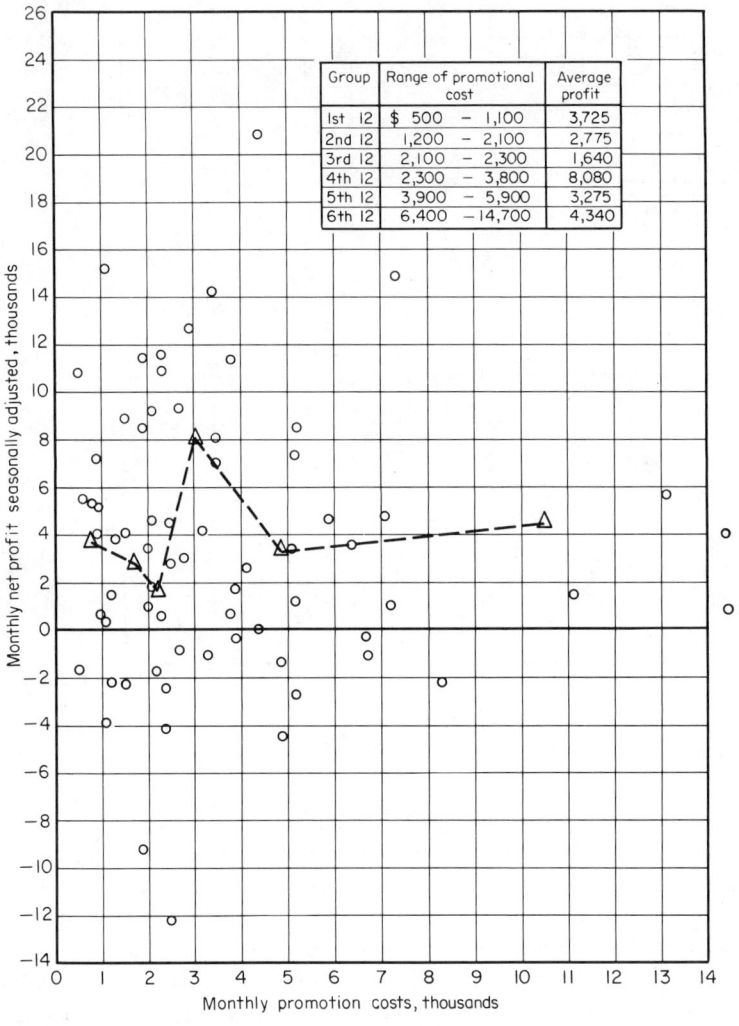

Fig. 3-3 Net profit vs. promotion cost (monthly).

BUSINESS STRATEGIES FOR INCREASED PROFIT

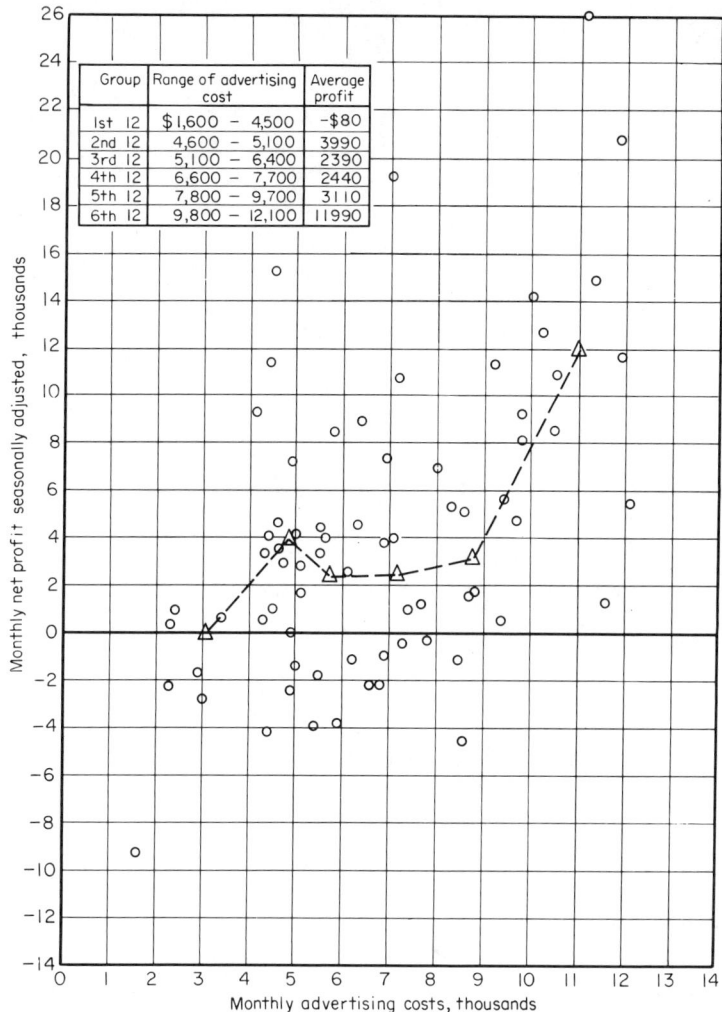

Fig. 3-4 Net profit vs. advertising cost (monthly).

expenditures in Fig. 3-3. This confirms the absence of overall upward trend, and in fact hints that some intermediate level of promotional expenditures may be the best bet. On the whole, promotions show little promise as a profitable business strategy for you.

Figure 3-2 shows advertising to be a different matter, despite a rather wide scatter of points that tells you a given advertising expenditure does not always result in the same profit outcome. Generally, higher profit has occurred in months of higher advertising effort, and other things being equal, you'd expect this trend to continue in the future.

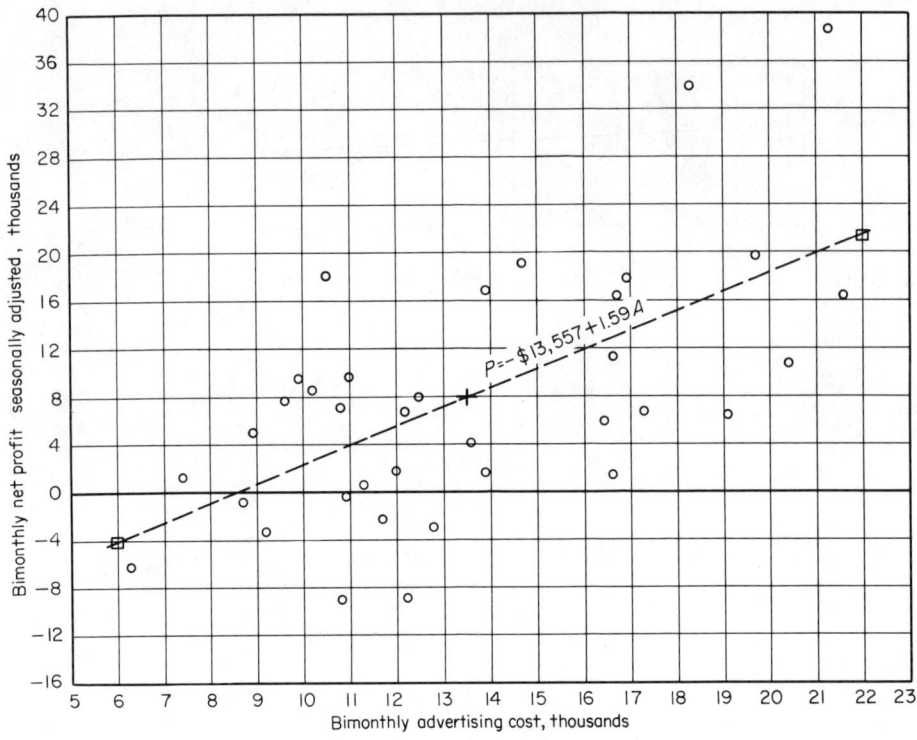

Fig. 3-5 Net profit vs. advertising cost (bimonthly).

When you plot averages by groups of 12 as you did before, in Fig. 3-4, a correlation shows clearly.

In some businesses, advertising may carry its effect into the next month or longer; in others its effect is transitory. Perhaps a combination of carry-over and immediate return fits your company best, so you try grouping advertising costs and profit into two-month units; for January-February, 1961, advertising costs would be (from Table 3-1) $4,500 + $8,000, or $12,500, and net profit would be $1,000 + $7,000, or $8,000—and all other two-month values would be computed similarly. Figure 3-5 plots the 36 two-month points, and now you can detect an unmistakable correlation. You are ready to analyze the relationship more precisely.*

* The beauty of using the swift and inexpensive computer is that you don't have to decide which of several possible relationships applies best. You try all those you think might apply, let the computer select the one that correlates best, and use that one.

REGRESSION ANALYSIS

You see that more advertising tends to produce higher profits, but you want your data to be more specific than this. You want to know:

What a dollar of advertising will produce in additional profits, everything else being equal

How accurate this prediction is, and how much confidence you can attach to it (or what odds you'd lay on it)

How do you guard against coincidences that might give a spurious cause-and-effect indication

Profit from a Dollar of Advertising

Statistical theory can find the one "best-fitting" line through the 36 points of Fig. 3-5—best-fitting in the sense that, of all lines you might draw to represent the trend of the points, it portrays best the average relationship between advertising costs and profits. (It is the one for which the sum of all the distances from each point to the line is the smallest.*) Calculations for obtaining the values needed to draw the line give the following equation for predicting profit based on level of advertising:

$$P_{sa} = -\$13{,}557 + 1.59A$$

This equation gives you an estimate of net profit for any two-month period (P_{sa}) by entering your planned advertising expenditure for that two-month period (A) into the equation and seeing what you get for P_{sa}. The line represented by this equation is plotted on Fig. 3-5.

Remember that the equation estimates seasonally adjusted profit, and you must deseasonalize it by reversing the process. Earlier in this chapter you calculated the seasonal adjustment factors for November and December as 144 and 60 percent, respectively, and therefore if you group these two months you must deseasonalize by using a factor that combines these two values. It is not a simple arithmetic mean, however, because it must give more weight to the lower value (corresponding to the greater effect the lower necessarily will have in the final summation); you use the "harmonic mean," which gives a combined factor of 85%.† Remember, too, that you added $20,000 before you seasonalized, to avoid the complications of negative values. After you estimate P_{sa} for November-December from the formula, then to get the actual (deseasonalized)

---

* Actually, you square each distance before you sum; in statistical parlance, this is the "least-squares" method. See B. E. Goetz, *Quantitative Methods: A Survey and Guide for Managers*, McGraw-Hill Book Company, New York, 1965, p. 251.

† The calculation is $2/(1/1.44 + 1/0.60) = 0.85$.

estimate of November-December profits, you would

> Increase P_{sa} by \$20,000 × 2, or \$40,000.
> Multiply by the November-December factor of 0.85.
> Decrease the result by \$40,000.

This estimating procedure applies only within the general range of past operations. When you try to estimate from it the expected profits from advertising at a level far above what you have ever tried before, the dependability of your estimate naturally drops off.

Let's consider this matter of dependability.

Accuracy and Dependability of the Prediction

You can determine the accuracy of your prediction for any confidence you wish, but the higher odds you demand that you are right, the less precise you can be. (You might give 9-to-1 odds that you could guess tomorrow's noontime temperature within a 40° range, but you couldn't be nearly so confident of hitting within 10°.) Statistical theory tells you how likely it is that your *calculations* of profit will lie within some specified dollar range of the *true* net profit. It does this mathematically by analyzing the scatter of your past points around the line on Fig. 3-5 and deciding how dependable the line is for future forecasting. This matter of confidence was discussed in Chap. 2.

Suppose you want your prediction to a 90 percent confidence; that is, when you predict a range of dollar values for average net profit that will be produced by any future level of advertising expenditure, you expect that the true average will fall inside this range 9 times out of 10. The least-squares computation produces the values needed to determine the width of this "confidence range." This range is the area enclosed by the pair of solid lines on each side of the "regression line" of Fig. 3-6.

But Fig. 3-6 also shows a pair of dotted lines enclosing a much wider "90 percent confidence range." What are they for?

The smaller range within the solid lines applies to future *average* values. If you were to advertise at some certain level for a dozen or two different two-month periods (not necessarily consecutive), getting various values of net profit for these periods, the *average* of these different values would be 9 times more likely to be inside this smaller range than not. The wider range within the dotted lines is that within which you expect 90 percent of the different *individual* two-month profit figures to fall. It is far harder to predict an individual value than an average one; but it is the average profit for a certain strategy that counts. Except in special circumstances (say a businessman wants an estimate of the worst profit he could reasonably make next month for a given advertising level), only the range enclosed by the solid lines is of interest.

BUSINESS STRATEGIES FOR INCREASED PROFIT

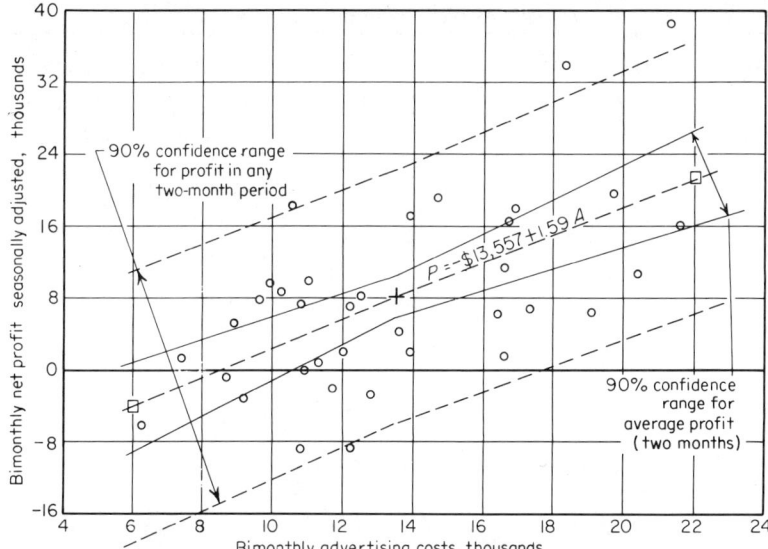

Fig. 3-6 Net profit vs. advertising cost (bimonthly).

DISTURBING FACTORS

Any cause-and-effect analysis faces the possibility that the relationship you find is not causal, but is an unfortunate combination of other factors that gives a false impression of causation. Most successful men have good vocabularies, but you are skeptical of advertised hints that increasing your "word power" *produces* success; you think a third factor—intelligence—tends to produce both success and literacy. High correlation between smoking frequency and lung cancer does not show conclusively that one causes the other; conceivably some common factor—say a high metabolism rate—could stimulate the smoking habit and independently could make people more susceptible to lung disease. You have to sniff out any such hidden factors, and see if they could have produced the results you're seeing. In the paragraphs that follow, three possible disturbing factors are considered.

1. *Perhaps all months of high advertising expenditure fell in certain seasons*, so that you really don't know what effect increased advertising would have in other seasons. Table 3-3 arranges various advertising expenditure levels by calendar months and calendar two-month periods, for the entire six-year sample. You are relieved to see a good mixture of high and low expenditures

Table 3-3   Distribution of advertising costs by months

Monthly	Jan.	Feb.	Mar.	Apr.	May	June	July	Aug.	Sept.	Oct.	Nov.	Dec.
Below 4,500	3	1	1				2	1	1			6
4,500– 5,500	1		2		1	4		2	1	1	1	
5,500– 6,500		1	2	2	2				1			
6,500– 7,500		1		2	1	1	1	2	1		1	1
7,500– 8,500	2	1										1
8,500– 9,500		1		1				1		2	1	2
9,500–10,500			2		1	1	1	1		1		
Over 10,500		1			1		1				1	2
Two-month grouping:												
Below 9,000	1		1			2						
9,000–11,000	1		2		2					1		3
11,000–13,000	2				2		1		2			
13,500–15,000				1								3
15,000–17,000	1	1			1			2				
Over 17,000	1	1			1			1		3		

throughout the year, and you conclude that this factor does not mask data and thus prevent it from being useful for prediction purposes.*

2. *Perhaps high advertising and high promotional expenditures correlate closely*, meaning that you've advertised heavily only when you had promotions to advertise, with the result that you really don't know what the effect of increased advertising would be in low-promotion months. Figure 3-7 divides each type of expenditure—promotional and advertising—into a high, a medium, and a low group, and arranges the 36 periods according to which of the resulting nine "cells" they fit.

If there were no association between advertising and promotion costs, you would expect a random distribution with approximately four outcomes in each cell:† four periods with high promotion and medium advertising, four with high promotion and low advertising, and so on. Such an "expected" distribution of outcomes by cell would be as in Table 3-4.

If you found high promotion and advertising costs always

---

* A company's best source of information for management decision is its own past performance *if* operations have been diverse enough to provide useful experimentation. The previously described technique of evolutionary operation is based on making small systematic changes in business strategy so that helpful information will be generated.

† You would not look for exactly four in each cell; this would be almost suspiciously good. If you throw a coin 10 times you know 5 heads is the most likely single outcome; but if you threw 10 coins thousands of times, you'd get an outcome of *exactly* 5 *heads* less than 1 time in 4. It isn't often that the "expected value" is more likely to occur than all other possible outcomes combined.

BUSINESS STRATEGIES FOR INCREASED PROFIT

	Two-month advertising cost			Average profit from promotion strategy
	Bottom third $6,300–10,900	Middle third $11,000–14,700	Top third $16,400–21,600	
Bottom third $1,400–4,700	$9,600 18,200 (−) 6,300 (−) 3,200 (−) 800 (−) 6,300	$6,800	$11,400 6,700 18,000 5,900 10,800	$5,900
Middle third $4,700–7,600	$8,600 5,100 (−) 100	$9,800 (−) 8,900 1,800 4,200 19,100	$16,500 6,400 19,800 38,800	$10,100
Top third $8,400–15,400	$7,100 1,300 7,700	$8,000 (−) 2,200 (−) 2,900 600 1,700 20,900	$1,500 16,400 34,100	$7,850
Average profit from advertising strategy	$3,410	$4,910	$15,525	

(Row label at left: Two month promotion costs)

Fig. 3-7 Two-month promotion and advertising costs.

going together, low always together, etc., you would have a distribution of outcomes by cells as in Table 3-5.

If you tallied the population of each cell from Fig. 3-7, you would find the actual distribution as in Table 3-6.

High promotions and high advertising do not correlate. Since there is a good random scatter, far closer to the "all 4s" than to the "12-12-12," this factor cannot produce a spurious correlation between advertising and profits.

Table 3-4

		Advertising costs		
		Low 1/3	Medium 1/3	Top 1/3
Promotion costs	Low 1/3	4	4	4
	Med. 1/3	4	4	4
	Top 1/3	4	4	4

Table 3-5

Promotion costs / Advertising costs

	Low 1/3	Medium 1/3	Top 1/3
Low 1/3	12		
Med. 1/3		12	
Top 1/3			12

3. *Perhaps your high advertising has occurred only when the economy is booming,* so that you cannot separate the effect on profits of advertising and economic forces. From the many available local business indicators,* you select a weighted composite index of business activity such as a university business school might compile. Since the economy changes relatively slowly, you can use yearly averages. Calculate the total profit, advertising cost, and promotional cost for each year; and compute the average value of the business activity index for each year. The values are graphed in Fig. 3-8. Initial inspection suggests a correlation between advertising expenditures and business activity, but closer observation discloses marked differences. In 1962, business activity rose and advertising decreased. In 1963, each reversed its direction. In 1965 and 1966, business activity doubled its rate of rise, from 4 to 8 percent, while advertising

* Activities compiling indicators of various sorts include Commerce Department field offices, Federal Reserve District Banks, universities, banks, state departments of labor and industry, employment offices, Dun & Bradstreet, post offices, utilities, trade associations, and many others.

Table 3-6

Promotion costs / Advertising costs

	Low 1/3	Medium 1/3	Top 1/3
Low 1/3	6	1	5
Med. 1/3	3	5	4
Top 1/3	3	6	3

BUSINESS STRATEGIES FOR INCREASED PROFIT

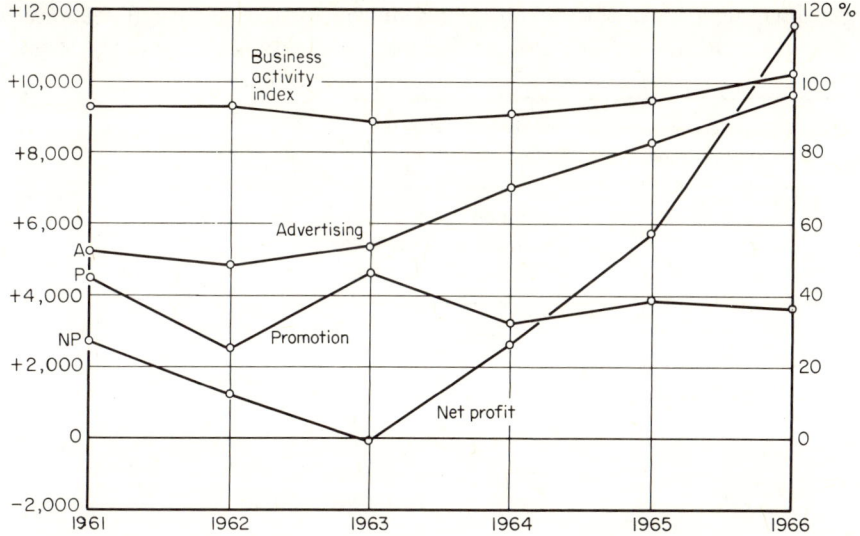

Fig. 3-8

costs maintained a steady climb. The differences are sufficient to make it clear that advertising policy took a course of its own, rather than simply mirroring the economy.

4. *Other factors:* The previous three factors may not be the only ones you should investigate. Your business may be affected by special activities—carnivals, area promotions, conventions, trade fairs, operations of competing firms—and you must make sure that your expenditures for the business strategy you are investigating have not risen and fallen in close correspondence with any of these.

CONCLUSIONS ON INTERNAL STRATEGY

This analysis suggests strongly that increased advertising will increase net profits; apparently your advertising budget is not close to the saturation level. Increased promotional activity, on the other hand, promises no improvement in net profits; in fact, both Fig. 3-5 and Fig. 3-7 suggest that some intermediate level may be better than a higher or lower budget. You are well advised to keep your predictions within the range of advertising expenditures bounded by your highest and lowest past expenditures. Figure 3-6 shows that predictive accuracy becomes progressively worse as you near the ends of this region. Within these limits, however, you can be 90 percent confident (reading from Fig. 3-6) that the following two-month advertising expenditures will produce net profits within the ranges shown, assuming normal economic activity (and remembering that these are seasonally adjusted net profits):

Advertising expenditure for 2-month period	Average net profit for 2-month period
$ 8,500	−$ 4,000 − +$ 4,000
13,500	+$ 5,600 − +$10,400
18,500	+$12,000 − +$20,000

4

Business Climate and Profit

INTRODUCTION

Purpose of this chapter

The previous chapter studied the effects of two different marketing strategies on net profit, discarded one of them because it has little effect on profit, and analyzed the other in detail to determine its effect on average profit in specific terms.* The shortcoming of this prediction is its neglect of the business climate in which you operate. When business is booming and the public optimistic, a certain level of advertising will appear to produce more profits than it really does—the economy should get the credit for the rest. If you can separate the contributions of advertising and business conditions, your predictions will be more dependable. This chapter will accomplish such a separation.

Plan of action

The following sequence of steps will achieve the prediction you require:
    1. Select the most representative barometer of business activity for your area.

  * If you had found two or more factors that correlated with amount of profit in the way advertising did, you could have calculated a prediction equation incorporating all of them.

2. Calculate how average net profit has varied with this business index in the past—and therefore should vary in the future—*ignoring variations in advertising expense*. (This is the exact opposite of what you did in the previous chapter.)
3. Step 2 has generated a prediction line showing expected profit for any value of the business index, assuming that advertising expenditure is at its average past value. Using this expected profit as a base, you now calculate the value of expected profit above or below this base which will result from advertising expenditures above or below their past average.
4. When you combine step 3 with step 2, you will have a predicted value of net profit for various values of advertising cost *and* business index—which is what you want.

SELECTION OF BUSINESS INDICATORS

Typical indicators available

National Indicators

Although the number and diversity of nationwide economic indicators are much higher than for regional or metropolitan indicators, the vastness of the United States makes them too general for most prediction purposes. They fall under the following general headings:
1. Production
2. Trade (including carloadings)
3. Prices
4. Financial
5. Employment
6. Construction
7. Governmental indices derived from the above

Local Indicators

A typical metropolitan area has the following indicators:
1. Production (6 subheads, chief of which is primary metals)
2. Trade (7 subheads, chief of which is retail trade)
3. Prices (6 subheads, 1 monthly and 5 quarterly)
4. Financial (5 subheads, all banking activity)
5. Employment (17 manufacturing subheads, 10 nonmanufacturing subheads, plus 2 unemployment subheads)
6. Freight (11 subheads)
7. Real estate (4 subheads)
8. General business activity index (university business school)

BUSINESS CLIMATE AND PROFIT

Selection of suitable indicators

The initial selection of promising indicators must be made on a judgmental basis, after considering the extent to which they appear related to your product and other factors. Assume that your business is retail sales of furniture and appliances, and make a quick review of the indicators available above.

Production indices should be promising for a manufacturing region; but hours worked in manufacturing measures the same business climate at a somewhat earlier date, since people must be employed before goods are delivered. Trade indices are important to you, particularly retail trade of firms like yours; food trade would not be as good an indicator. Telephone installations might seem promising, as might real estate transactions; but the former show a long-term rise in good times and bad, and the latter are too inaccurate and too delayed. Freight shipments are tied closely to manufacturing indices. Car registrations, utility sales, postal shipments, and similar measures tend to be "following" indicators which are not as useful for heralding future economic trends. The general business activity index, being a weighted compilation of most other measures, comes out somewhat late and thus is less useful as a leading indicator.

After due consideration, you decide to plot these six indicators:
1. General business activity index
2. Unemployment rate
3. Hours worked, all manufacturing
4. Hours worked, primary metals
5. Department store index
6. Bank clearings

The average annual value of each index is plotted on Fig. 4-1 against the average annual value of net profit, seasonally adjusted. It can be seen from these preliminary plots that *Unemployment rate*, *Hours worked, all manufacturing*, and *Bank clearings* have reasonably good correlation with your net profit, so you select these three to examine in more detail.

ANALYSIS OF SELECTED INDICATORS

Purpose of analysis

What are you trying to do with these economic indicators? For one thing, you want to separate the influence of business conditions from the influence of advertising, so you can determine more accurately the

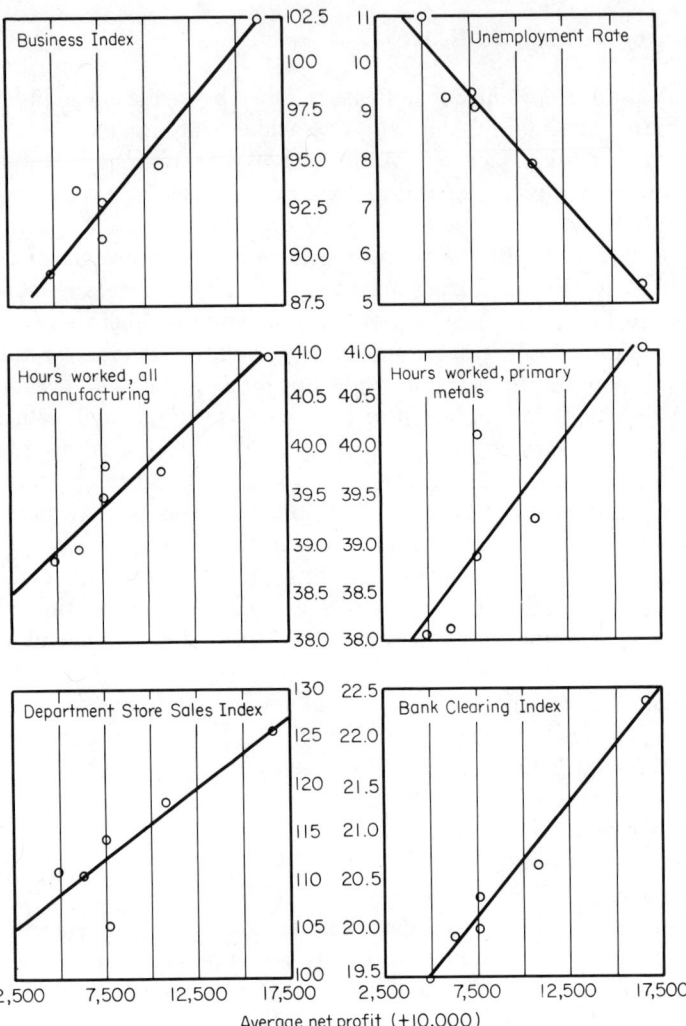

Fig. 4-1 Various indicators vs. net profit (1961–1966).

individual effect of the latter. Additionally, you want to predict the future. It doesn't help you to know what effect January's business climate has on your January sales; you need to know what it will do to your March or April sales. You compare these three potential indicators, then, on the basis of how well they have correlated, in the past, with your profit performance two months later.

For each of the three indices selected, you will need to relate the value for each month with the value of net profit for *two months later*. For

BUSINESS CLIMATE AND PROFIT

example, the unemployment rate for January, 1961 (11.5 percent), must be related to the net profit for March, 1961 ($18,600).

Unemployment rate versus profit

The first indicator analyzed is unemployment rate. Figure 4-1 has shown you that there is correlation as far as annual averages are concerned; now you will determine just how close that correlation has been on a month-by-month basis, *using unemployment rate as a two-month leading indicator.*

You could go through all the same process you used in Chap. 3 when you were determining the effect of advertising on profit: plot each monthly value,* note the existence of a trend by eye, calculate the best-fitting line and draw it on the graph, and calculate and plot the range for a "90 percent confidence" prediction of average profit given the level of advertising (or unemployment rate in this case). There's no need for all this. You want to know which of the three potential indicators has the best correlation with profit—that is, has the tightest 90 percent confidence range—and a statistician using a least-squares computer routine will give you this for all three indicators in half an hour.

A computer solution for unemployment rate versus profit produces the following equation for the relationship:

$$P_m = \$17{,}360 - \$1{,}526 \times U_{m-2}$$

where P_m is the predicted average profit for any month and U_{m-2} is the percentage of unemployment rate for two months earlier.

The solution also tells you that the 90 percent confidence band at its smallest point (the point where most of your experience has been, and thus where most of your plotted values congregate) is $\pm\$1{,}155$. At this point then, the average value of profit predicted by the above equation, solely on the basis of the unemployment rate two months earlier, will be within $1,155 of the true average value 9 times out of 10.

Hours worked, all manufacturing, versus profit

A similar calculation undertaken for this indicator produces the following equation:

$$P_m = -105{,}970 + 2{,}740 \times H_{m-2}$$

The 90 percent confidence band at its smallest point is $\pm\$1{,}370$. Since this is larger than the band for unemployment rate, this indicator turns out to be less reliable an economic predictor than unemployment rate.

* You wouldn't combine into two-month values as you did in advertising, since you are trying to test the value of a *monthly* economic indicator.

Bank clearings versus profit

A third calculation for this indicator produces the following equation:

$$P_m = -21{,}645 + 1{,}260 \times C_{m-2}$$

and a 90 percent confidence band at its smallest point of $\pm\$1{,}265$. This indicator too is rejected. Unemployment rate is the most dependable predictor of net profit.

Further analysis of unemployment rate

Perhaps unemployment rate can predict net profit with reasonable precision even further into the future. This is tested by comparing net profit with unemployment three months previous. The results of this calculation are:

$$P_m = 16{,}910 - 1{,}453 \times U_{m-3}$$

The 90 percent band at its smallest point is $\pm\$1{,}193$. The confidence band is only slightly larger than that for U_{m-2}, and it still permits prediction with reasonably good accuracy.

COMBINING INTERNAL AND EXTERNAL STRATEGIES

Concept

The problem of predicting net profit has been approached in two ways:
1. By seeing how it varies with level of advertising, assuming the external economy does not change appreciably.
2. By seeing how it varies with business conditions, assuming no significant change in level of advertising.

Neither of these assumptions is warranted, for the previous calculations have shown clearly that both advertising and business conditions vary, and that net profit is affected by the variation of each.* It is time to combine these two factors into a single predictive equation.

The rationale of this process is as follows. Every month for the

---

* Many other things affect net profit, but to a lesser degree. Since operations research models must be simplified versions of reality, if they are to avo'd hopeless complexity, these other factors have to be ignored. Their combined effect is the principal ingredient which gives your predictions the form of confidence ranges rather than precise lines. If you find your ranges of error unacceptably large, you have to corral more causative factors and include them in your calculations.

BUSINESS CLIMATE AND PROFIT

past six years you made some *actual* net profit. You decided that unemployment rate had a good deal to do with the amount of profit, and you developed a prediction equation:

$$P_m = \$17,360 - \$1,526 \times U_{m-2}$$

which enables you to calculate the *expected* net profit you would get if unemployment rate were the only variable. In March, 1961, your *actual* net profit was $18,600 (seasonally adjusted, and with a dummy $20,000 added); and unemployment rate (for January, 1961, since you are using it as a two-month leading indicator) was 11.5 percent. Your expected net profit from the above equation (after adding the dummy $20,000) would be $19,809. The difference between $19,809 and $18,600 is caused by less than usual advertising for this month.

Now to use this concept. You are assuming that some imaginary base profit level would be produced if business conditions and advertising effort remained constant at some average level. You apply the upward or downward effect of business conditions to this—from the above equation for P_m—and you get an "expected profit" as it might vary with the economy, assuming advertising to remain constant at its average level. Now using this expected profit as a base, you compute another regression line showing how net profit might go up or down with advertising *from this expected profit base*. This is a two-step process, but at the end you can combine both steps into one prediction equation.

You've accomplished the first half of this two-step process in deriving the above equation for P_m: insert the unemployment rate for each of the 70 months, and solve the equation 70 times for expected net profit.* If you subtract each of these *expected* values from the *actual* value for that month, you have the *increment of profit produced by advertising*. You have, in other words, 70 values showing what advertising produced each month; and if you calculate another regression-line equation you will have a mathematical expression showing the exact way in which a certain level of advertising produces this increase or decrease in profit over what is produced by business conditions alone.

The separate effects of advertising

Profit Increase from Advertising

You have seen above that, if you use the prediction equation based on unemployment rate, the *expected* March, 1961, profit (predicted on the basis of economic conditions alone, and assuming average advertising expenditures) is calculated by inserting the value of 11.5 percent unemployment for January, 1961, and has a calculated value of $19,800.

---

* You don't actually solve the equation 70 times; the computer does it for you.

Since the *actual* March, 1961, profit is $18,600, the March 1961 profit increment produced by the *separate* effect of advertising is $18,600 − $19,800 or −$1,200.

For April, 1961, similar calculations would give a profit increment produced by the separate effect of advertisement of $28,500 − $20,400 or $8,100.

Since you concluded in Chap. 3 that advertising produced a carry-over effect, you grouped advertising and profit into two-month values before calculating your prediction equation. The two-month profit increment for March-April, 1961, produced by advertising would be (in hundreds of dollars):

$$(\text{Actual profit} - \text{expected profit})_{\text{March}} = \$186 - \$198 = -\$12$$
$$+ (\text{actual profit} - \text{expected profit})_{\text{April}} = \$285 - \$204 = +\$81$$
$$= (\text{profit increment from advertising})_{\text{March-April}} = \underline{+\$69}$$

The values of this bimonthly profit increment from advertising for all bimonthly periods and the corresponding bimonthly advertising expenditures are plotted together in Fig. 4-2. The points show a general upward trend of profits with advertising level; and the plot of averages for groups of six (the triangular points) shows a clear correlation, though by no means a smooth one.*

---

* An exact correlation is too much to hope for in the real world of marketing. If the relationships were as clear as all that, there would be no need for quantitative aids. It must be remembered, however, that any correlation at all constitutes very valuable intelligence to the decision-maker.

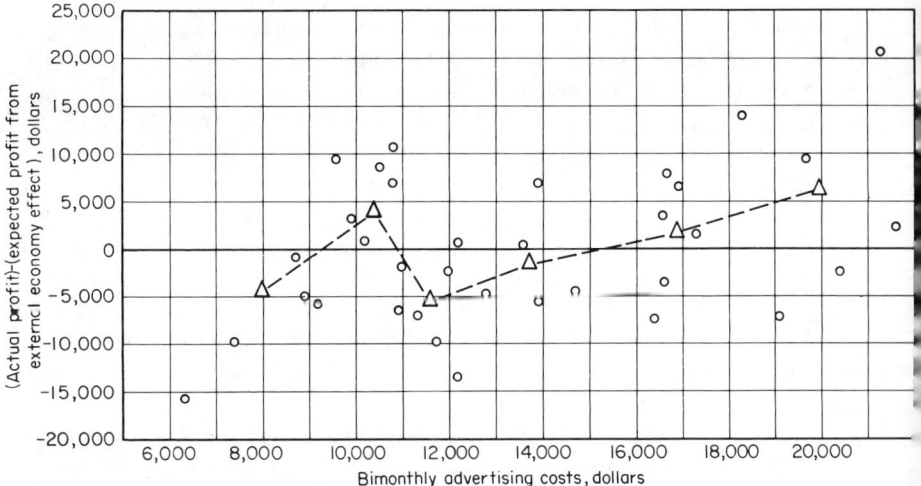

Fig. 4-2 Incremental profit vs. advertising. Expected profit from economic conditions vs. base (bimonthly).

BUSINESS CLIMATE AND PROFIT

A prediction equation for the relationship shown on Fig. 4-2, calculated as before, would give the following equation and minimum confidence interval:

$$P - E(P)_{econ} = -\$10,700 + 0.79 \times A$$

where $E(P)_{econ}$ is the expected profit based on economic conditions. The minimum confidence interval for 90 percent confidence is $\pm\$2,076$.

The combined prediction equation

Your final step is to combine these two prediction concepts you have developed—expected profit arising from business conditions under the assumption that your advertising level is constant, and expected increase or decrease from this amount arising from an advertising level higher or lower than usual—into a single prediction equation for your use.

Suppose it is early April, and you have just got the unemployment rate for March. You want to predict your bimonthly profit for May and June.

Expected profit from economic climate for May (two months later than March) would be

$$E(P)_{May} = \$17,360 - \$1,526 \times U_{March}$$

Expected profit from economic climate for June (three months later than March) would be

$$E(P)_{June} = \$16,910 - \$1,453 \times U_{March}$$

Expected profit from economic climate for May and June combined would be simply the sum of these:

$$\begin{aligned}E(P)_{May\text{-}June} &= \$17,360 - \$1,526 U_{March} + \$16,910 - \$1,453 U_{March} \\ &= \$34,270 - 2,980 U_{March}\end{aligned}$$

When you combine this equation with the prediction equation you calculated above for the separate effect of advertising, you have:

$$P_{May\text{-}June} - (34,270 - 2,980 U_{March}) = -10,700 + 0.79 \times A_{May\text{-}June}$$
or $P_{May\text{-}June} = 23,570 - 2,980 U_{March} + 0.79 A_{May\text{-}June}$

This is your final prediction equation. It says that the net profit (seasonally adjusted) for any two-month period can be estimated by inserting into the above equation the unemployment percentage for two months earlier than the start of the period and the anticipated advertising level for the period. What is the accuracy of this estimate for predicting *average* two-month profits? Its combined minimum confidence interval for 90 percent confidence (and remember that it must take care of all errors contributed by both unemployment and advertising) is ±$2,680.*

USE OF THE PREDICTION EQUATION

The equation will predict, with 90 percent confidence, what profit you will average in the long run from a combination of certain economic conditions and certain advertising expenditures. In one two-month period your profits might be above the prediction range, in another they might be below—but when you average 10 or 12 such two-month periods, this average profit will fall within your confidence range 9 times out of 10.

You must remember that the profit predicted by the equation is seasonally adjusted. You must deseasonalize it before you can use it as a valid estimate of your profit for any two-month period. If you are predicting net profit for May-June, then strictly speaking you should multiply the May part of predicted profits by May's seasonal adjustment factor from Fig. 4-1, and multiply the June part by June's seasonal adjustment factor. Since you have no way of knowing how your predicted May-June profit divides, you must average the seasonal adjustment factors for May and June (using the harmonic mean; page 49), and multiply the total predicted profit by this.

The highest two-month advertising budget your business has tried is around $21,000. Since your prediction equation loses precision fast when it gets much outside the range of previous data, you cannot use the findings of this chapter to justify your adoption of an advertising strategy that *averages* much higher than this. Individual two-month periods may go higher and lower, and indeed they should if you are to develop new operating experience to enrich your store of managerial decision data.

SUMMARY

What have you done in this chapter?

You considered the elements of marketing strategy that apply to your business and that can be costed separately—this chapter considered only two but it could have reviewed as many as necessary—and you decided which showed some consistent relationship to profits.

* The standard deviation, or measure of error of an equation which is the sum of separate expressions, is the square root of the sum of the squared standard deviations of the separate expressions.

BUSINESS CLIMATE AND PROFIT 67

You selected the one* that correlated significantly with profit and developed a prediction equation, disregarding external economic factors.

You extended your prediction equation to take account of external business conditions by considering a number of business indicators and selecting the one that correlated best with your past profit performance.

The end result of this exercise was a "prediction equation" for net profits. At best, this equation tells you within certain limits and to a certain accuracy what average net profit you can expect from various combinations of uncontrollable marketing strategy. At worst, it gives you a far better understanding of the relative effect of various factors on profits—a sort of sensitivity analysis for each factor.

Your business is far more complex than this? Many more factors are operating? No matter. The concept of this chapter holds up well under increased complexity—it is only the mechanics that vary to take account of additional disturbing factors. You should remember this: the more complex your operations, the more indispensable these quantitative techniques that enable you to peer through the murk and discern whatever order lies within.

* You could have selected two or more, with some increase in complexity but no change in concept. A little thought will tell you that two internal factors could be introduced one at a time, in the same way that this example introduced one internal and one external factor; you then could treat the expected profit from these two the way this example treated the expected profit from unemployment, and repeat the process to incorporate an external factor. If the process gets much more involved than this, you are well advised to use a multiple regression program.

5

Profitable Inventory Management

INTRODUCTION

The importance of inventory control

A very keen businessman telephoned me recently. Constantly on the outlook for new business methods, he had installed a computerized inventory-control system that was a whiz. For every item in his inventory it kept continuous tabs on stock level, selling rate, minimum and economical order quantity, replenishment rate, and profitability; and every morning it typed the day's purchase requirements based on its analysis of idealized inventory control.

The system had just one flaw: his stock level had swollen $200,000 while his gross sales were unchanged.

A standard element of inventory cost is interest at the usual bank rate on this tied-up capital; but if this needless $200,000 put my friend near the limit of his line of credit and thus foreclosed some profitable use of this capital, he was effectively paying a far higher rate. Another standard element of cost, additional warehousing, apparently did not apply since he was making his existing space do; but delays and overtime showed that he was feeling this element of cost, too. His computer control system hadn't led him to reduce his manual costs of inventory control, because—wise man!—he wasn't ready to trust it yet; so the computer cost was all additional.

This is not an argument against innovation.* It is a plea for giving this major element of controllable cost (inventory can tie up a fourth of total invested capital) the marginal attention it deserves. A recent engineering analysis of the relative economics of nuclear versus coal generating systems hinged on the depressingly mundane factor of interest on fuel inventory. In many cases this factor, lifted from the obscurity of incomplete analysis, will prove a heavier drag on profits than management suspects.

The economics of inventory

The conventional inventory account trades off the two costs of procuring and of stocking; the less you spend to carry a large stock, the more you must spend on frequent and thus costlier small procurements. Such an account omits a third cost fundamental to being in business: the cost of not having goods on hand when a buyer appears. The businessman obviously appreciates this factor, but by failing to treat it as analytically as other costs he risks giving it disproportionate attention—perhaps too much, perhaps too little.

Inventory management is a problem in economics—a search for the policy that maximizes the excess of receipts over costs. This chapter is concerned with systematic techniques to help you find such an inventory policy.

Contents of this chapter

You can look at inventory problems in two ways. First: How do you determine the cheapest way to procure and store material to meet a specified demand? Second: How, in the face of uncertain demand, do you identify the best compromise between being caught with more than you can sell and losing sales by running out of stock? This chapter first discusses techniques for finding an optimum inventory policy under various operating conditions, but assuming that demand is known; it then deals with the additional complications that arise when demand is uncertain.

* Though it certainly is a plug for planning before you leap. Some companies with computerized stock control and reorder functions say they could not remain competitive otherwise; but a major apparel chain which went bankrupt a few years ago had a grossly disoriented computer inventory system high on its roster of critical ills.

BASIC INVENTORY THEORY

Types of inventory costs

If you know that your business will require, for use or resale, one of an item daily for the coming year, your inventory policy can vary all the way from procuring a single item every day to procuring a year's supply at one time. The larger the one-time order quantity, the higher will be some costs and the lower others.

Costs that *increase* with order quantity are those involved with paying for and keeping items you will not need until later: interest on the tied-up inventory; obsolescence or deterioration of stock; inspection and maintenance; warehouse space rent; warehousing operations; taxes; insurance; theft; etc.

Costs that *decrease* with order quantity are those that benefit from large-scale operations: get-ready costs, which depend on number of manufacturing runs rather than the size of a run; cost of shutdowns from running out of stock; unit item costs, which usually drop as order quantity rises; labor costs (including supervision) which are lower with more level workloads associated with large production runs; costs of ordering, shipping, and receiving which may vary more with number of orders than with quantity ordered; etc.

The problem of the manager is to determine with reasonable accuracy what these costs are and how they change with order quantity, so that he can compute the inventory level at which the sum of all costs will be a minimum.

The simple inventory problem

Suppose you are facing an inventory problem involving *ordering costs that decrease* with size of order, and *holding costs that increase* with size of order. You need 100 items on the first of each month, at a unit cost of $5 an item. Your ordering costs (purchasing, invoicing, receiving, etc.) are $40 an order. Your holding costs consist of interest on cost of all items in stock at a rate of 10 percent a year, and storage at a rate of $1 a year for each item in stock. You receive and pay for incoming materials on the first of the month. Thus, if you buy annually, you receive 1,200 items on January 1 and use 100 items immediately; you will therefore pay storage on 1,100 items and interest on $5,500 until February 1. At that time you use 100 more items, and you will pay storage on 1,000 items and interest on $5,000 until March 1, and so on.

Table 5-1 shows, for different numbers of orders a year (in every case

PROFITABLE INVENTORY MANAGEMENT

Table 5-1

Orders per year	Items per order	Price per order	Interest on items stored	Storage on items stored	Holding costs (I + S)	Ordering costs	Total inventory costs
1	1,200	$6,000	$275	$550	$825	$ 40	$865
2	600	3,000	125	250	375	80	455
3	400	2,000	75	150	225	120	343
4	300	1,500	50	100	150	160	310
6	200	1,000	25	50	75	240	315
12	100	500	0	0	0	480	480

a total of 1,200 items is ordered for the year), the ordering cost, holding cost (interest plus storage), and total cost of inventory.

Of all the variations considered, four orders per year gives the lowest cost of carrying inventory. The difference between four and six orders per year is negligible, however, and either one could be considered an optimum choice.

Figure 5-1 portrays the same picture graphically. In a two-dimensional graph it may seem that the optimum cost region can be located more systematically than by tabulating all the possibilities as in Table 5-1. In real inventory problems, there usually are complicating factors which make it infeasible to resort to either of these simple solutions.

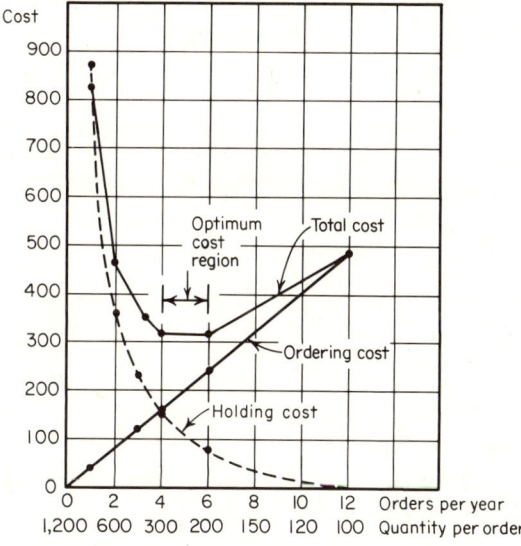

Fig. 5-1

Analytical solution

Either a trial-and-error or a graphical method will solve the problem, but both are too slow to be very practical. You need a method of solving for the optimum order quantity in a single step.

Express the significant costs and times of an inventory situation in generalized form, as follows:

S = ordering cost ("setup" cost, if a production run)
H = holding cost to store one item for one time unit (storage plus interest)
R = number of items needed per time unit
I = interval between orders (or production runs)
Q = size of each order (or production run)

In the previous example, these would have the following values:

S = $40 per order
H = $1 + 10% of $5 = $1.50 per item per year
R = 1,200 units per year
I and Q are the variables you seek to determine; they are related by the expression $R = Q/I$.

The amount produced each run, Q, is equal to RI, and the average inventory over any time period—if you assume that items are used up continuously at a constant rate—is $RI/2$. Figure 5-2 shows this situa-

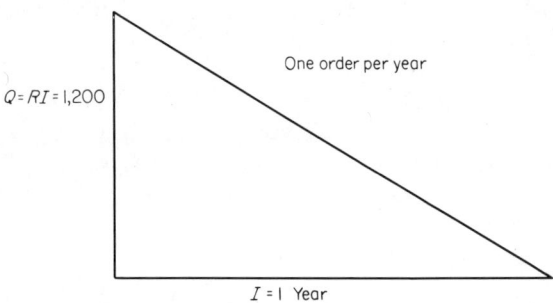

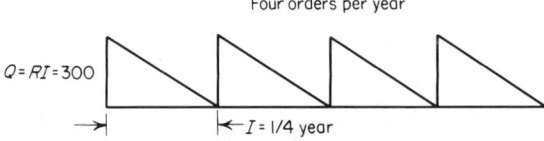

Fig. 5-2 Inventory replenishment and use—diagrammatic representation.

tion of periodic replenishment and constant usage for one order a year and for four orders a year; the average inventory is simply the area under the sawtooth triangle divided by the period. Since the cost of holding inventory per unit time is the holding cost times the average inventory, or $HRI/2$, and the ordering (or setup) cost per unit time is S/I, the total cost per unit time is the sum of these:

$$c = S/I + HRI/2$$

The minimum value of this expression for total cost per unit time* occurs when

$$I = \sqrt{\frac{2S}{HR}}$$

or (since Q, the amount produced each run, $= RI$) when $Q = \sqrt{2SR/H}$; the formula for *economic order quantity*.

Trying this on the previous example,

$$Q = \sqrt{\frac{2 \times 40 \times 1{,}200}{1.5}} = \sqrt{64{,}000} = 252\dagger$$

Extensions of the simple case

This analysis made several simplifying assumptions. It assumed that usage rate was constant, and was known perfectly in advance. It assumed that material would arrive exactly when planned, and that no ordering lead time was involved. It made no allowance for safety stocks (with supply and demand scheduled so precisely, why should it?) It took all demands to be absolute requirements, not weighing the implicit cost of being late in meeting an occasional demand against the cost of 100 percent punctuality. It did not consider the effect of quantity discounts from suppliers, or the possibility of meeting emergency needs by premium procurement methods. It did not treat the situation of production for simultaneous stocking and use. In the following sections, some of these complicating situations will be analyzed.

---

* Found by the calculus technique of "differentiating with respect to I" and setting the result equal to zero—a mathematical way of finding where, along the base line of Fig. 5-1, the total cost curve stops falling and momentarily attains zero slope before it starts rising.

† This differs slightly from the minimum-cost order of approximately 300 computed in the step-by-step solution, because this formula assumes that inventory is being used up continuously rather than in monthly increments at the start of each month.

Scheduling of Production Runs

Suppose that an order, instead of being filled at once and in full from stock (as in the previous case), is filled by scheduling a production run at some rate faster than the rate of use. The symbols for this situation will be:

S = setup cost
H = holding cost
R = usage (requirements) rate
I = interval between production runs
Q = quantity ordered each production run
P = production rate

This situation is shown in Fig. 5-3, with a production rate of 3 units a day and a usage rate of 1 unit a day. During the production-and-use phase, inventory rises at a rate of 2 units a day; during the use-only phase, inventory falls at the rate of 1 unit a day. Average inventory over the period is the area of the sawtooth triangle divided by its base. The base is I; the altitude is the inventory at the moment production stops, or $(P - R)I_p$. The average inventory is therefore $[(P - R)I_p]/2$, and the unit cost of holding inventory is H times this. Since the unit setup cost is S/I as before, the total cost per unit time is the sum of these, or $c = S/I + [H(P - R)I_p]/2$. To express both I_p and I in terms of Q, so that it can be differentiated as before, note that the amount produced equals the amount used in any cycle; therefore $PI_p = Q$, or $I_p = Q/P$, and similarly $RI = Q$, or $I = Q/R$. Substituting these in the cost equation and differentiating gives a minimum-cost production run of

$$Q = \sqrt{\frac{2SR}{H(1 - R/P)}} \quad \text{or} \quad I = \sqrt{\frac{2S}{RH(1 - R/P)}}$$

Comparing these with the previous results—for the case when procurement was instantaneous—you see that the optimum production run

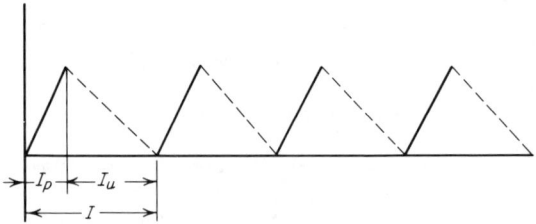

Fig. 5-3

PROFITABLE INVENTORY MANAGEMENT

and interval between runs now are larger by virtue of the fact that the denominator contains a factor:

$$1 - \frac{\text{usage rate}}{\text{production rate}}$$

and the nearer the usage rate approaches the production rate the larger will be the optimum production run. In the happy situation where production and usage just balance, the optimum production run goes on forever; but this utopian situation exists only in a plant forever free of breakdowns, strikes, or shortages.

Quantity Reduction in Orders

What will be the effect on optimum order quantity if the seller gives a price reduction for increased quantity of overall purchases per unit time—say the basic cost of an item is $5 but there is a reduction of item cost of $1/10$¢ times the total number ordered in a year? (For example, the item cost if 1,200 per year are purchased would be

$5 - 1,200 \times \$0.001 = \$3.80$ each.)

Let the unit cost be U, the reduction rate for quantity be r (so that the unit-cost reduction would be Rr), the interest rate per year be i, and the warehousing unit cost per year be W. Now the holding cost per item per unit time would be

$W + (U - Rr)i$

Grouping this into fixed and variable costs, it would be

$(W + Ui) - Rri$

and if the fixed component were expressed as "base holding rate" (holding rate if there were no quantity reduction, corresponding to the previous examples), and designated H', the total cost expression would be

$c = S/I + H'RI/2 - riR^2I/2$

The minimum-cost order quantity (differentiating) in this case would be

$$Q = \frac{2SR}{H' - 2riR}$$

The effect of a quantity reduction for annual purchases is, as expected, a decrease in holding costs and thus an increased optimum order quantity.

The problem if the seller gives a price reduction for increased quantity of each *individual* purchase is somewhat more complicated. This would be the case if there were, for example, a lower unit price for one 1,200-

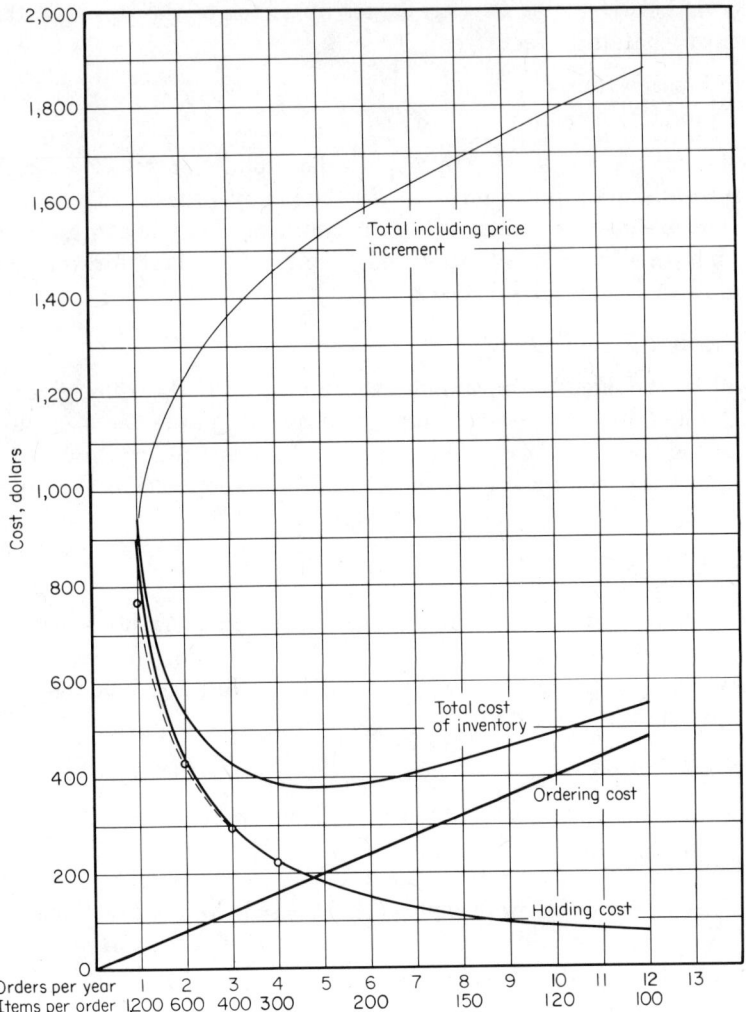

Fig. 5-4

item purchase than for two 600-item purchases. In this situation the total cost expression would be slightly different:

$$c = S/I + H'RI/2 - riR^2I^2/2$$

This leads to an expression for minimum value of Q which is not solved as easily as were the previous expressions:

$$2riQ^3 - H'Q^2 + 2SR = 0$$

PROFITABLE INVENTORY MANAGEMENT

Assume that the value of r is $0.001 per item in an individual order; thus an order of 100 at one time would save $0.001 × 100 or $0.10, for a unit price of $4.90; an order of 1,200 at one time would save $1.20, for a unit price of $3.80. The above equation, with values for the coefficients put in, would be

$$Q = \sqrt{\frac{96{,}000}{1.5}} = 252$$

If the equation is solved graphically, in the same manner as in Fig. 5-1, with and without a quantity discount based on run size, the results will be as shown in Fig. 5-4. The solid line for holding cost is without a discount, and the dotted line for holding cost is with the discount of the example above. The total-cost curve shown is for the no-discount case; but it can be seen that the minimum-cost order quantity of the total-cost curve will be at almost precisely the same point whether there is a discount or not. This discount rate is far from negligible, since it varies the unit-cost rate from $4.90 to $3.80, but its effect on *inventory* cost is hard to find.

The effect on total cost *including* purchase cost is, of course, substantial. In the example used, its effect overwhelms that of all other factors. The pertinent cost figures are tabulated in Table 5-2, and the cost curve showing inventory cost plus incremental purchase price difference is plotted at the top of Fig. 5-4. In another case, where quantity discounts are not so substantial or where they occur at specific volume levels rather than continuously, the outcome may be different. In any case, it will pay to analyze any proposition in this manner, to learn where the true optimum lies.

Table 5-2

Quantity per order, Q	Holding cost		Total inventory cost		Incremental purchase price, excess over $Q = 1{,}200$ rate	Total including price increment
	No discount	With discount	No discount	With discount		
100	$ 75.00	$ 74.50	$555.00	$554.50	$1,320.00	$1,974.50
120	90.00	89.28	490.00	489.28	1,296.00	1,785.28
150	112.50	111.38	332.50	431.38	1,260.00	1,691.38
200	150.00	148.00	390.00	388.00	1,200.00	1,588.00
300	225.00	220.50	385.00	380.50	1,080.00	1,460.50
400	300.00	292.00	420.00	412.00	960.00	1,372.00
600	450.00	432.00	530.00	512.00	720.00	1,232.00
1,200	900.00	768.00	940.00	808.00		808.00

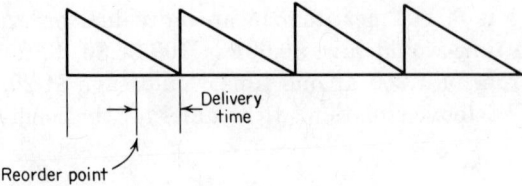

Fig. 5-5

Lead Times and Safety Stocks

It has been assumed thus far that orders are delivered at once. A somewhat more realistic assumption is that it takes a known specific time after an order is placed until the goods are received.

If usage is at a constant rate, as in Fig. 5-2, the solution is simply to backtrack from each required date a distance equal to the known delivery time, and place the order at that time. This situation is shown in Fig. 5-5. If usage rate varies in some predictable way, such as a seasonal variation, the delivery time is backtracked in the same way; the inventory level which triggers a reorder is shown for each period by the boldface arrows of Fig. 5-6.

Perhaps the delivery time is not constant, but you do know something from your records about how it varies. Suppose the supplier promises an average delivery time of 20 days, but your record of 32 past deliveries (which you believe to be typical of his usual practice) shows the following performance:

Delivery time from order date, days	Frequency with which this time has occurred
16	1
18	5
20	10
22	10
24	5
26	1

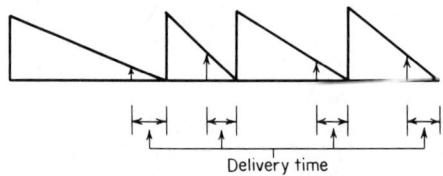

Fig. 5-6

If you plan on a 20-day delivery time, with no safety stocks to tide you over, you will run out of stock 50 percent of the time. If you play it safer, and plan on a 22-day delivery time, you will still have stockouts, but only 19 percent of the time. If you plan on a 24-day delivery, stockouts will occur only 3 percent of the time; and if you plan on a 26-day delivery you will be essentially free from stockouts.

Which should you do? As in all other economic business decisions, it is a matter of finding the minimum-cost compromise between the cost of running out and the cost of maintaining excess stock.

The cost of maintaining excess stock is computed in a straightforward way. If you order with 16 days lead time, you will never have to store any excess inventory, but you will run out of stock 31 times out of every 32, for periods varying from 2 to 10 days at a time. If you order with any more lead time than this, you will incur holding costs part of the time. Assuming a fixed holding cost of $1.50 per item per year, as before, and assuming that you order 300 items 4 times a year (the optimum ordering pattern calculated previously), for each day that an order arrives ahead of the desired date your holding cost will be ($1.50 × 300)/365 or $1.23. If your lead time is 20 days, for example, then out of each 32 shipments there will be 5 (16 percent) that arrive in 18 days and have to be held 2 excess days, and there will be 1 (3 percent) that arrives in 16 days and has to be held 4 excess days.

You could evaluate this by saying that 16 percent of the time you will have holding costs of $1.23 × 2, 3 percent of the time you will have holding costs of $1.23 × 4, and the rest of the time you will have no holding costs. You would reach the same conclusion by computing the average, or "expected," holding cost per order as follows:

$$\text{Expected holding cost} = \$1.23[2 \times 16\% + 4 \times 3\%]$$
$$= \$1.23[0.32 + 0.12] = \$0.54$$

Calculating holding costs for all lead times will give the results shown in Table 5-3.

So much for the costs of holding excess stocks. In this example they are modest, and perhaps you would be tempted to play it completely safe and settle for a 26-day lead time for all orders. This is not a rational decision, because another alternative may have lower total cost. In order to explore this, you must make an effort to assess the cost of stockouts—and this is not so straightforward. The second part of this chapter evaluates the "opportunity loss" of not having stock on hand when there is a market for it (a market that is not perfectly predictable but about whose pattern you can get some information). If a stockout requires you to make alternate arrangements, such as buying items at a premium price to tide you over, using higher-cost production methods, or putting work-

Table 5-3

Order lead time, days	Holding cost per item per day	Expected no. of item-days per order	Holding cost per order	Holding cost per year
16	$1.23	0	$ .00	$ .00
18	1.23	0.06	.07	.28
20	1.23	0.44	.54	2.16
22	1.23	1.44	1.77	7.08
24	1.23	2.06	3.76	15.04
26	1.23	5.00	6.15	24.60

men on less productive work until the stockout ends, you can estimate such costs with fair accuracy. Suppose you have established somehow that stockout cost varies with the number of item-days of stockout (a shortage of 10 items for 2 days is equivalent to a shortage of 5 items for 4 days), and that the variation is linear (a shortage of 10 items for 2 days costs twice as much as a shortage of 10 items for 1 day).

If you adopt a 26-day lead time, you will virtually never run out of stock. If you adopt a 22-day lead time, you will have a stockout lasting 2 days 16 percent of the time and a stockout lasting 4 days 3 percent of the time. Since your consumption rate is 3.3 items a day,* this amounts to an expected number of stockout days as follows:

If 2 days late:
$$3.3 \text{ for 2 days} + 3.3 \text{ for 1 day} = 9.9 \times 16\% = 1.58$$

If 4 days late:
$$3.3 \times 4 + 3.3 \times 3 + 3.3 \times 2 + 3.3 \times 1 = 33 \times 3\% = \frac{0.99}{2.57}$$

In general, if the order is N days late, the number of stockout days will

* 1,200/365, ignoring the complications of weekends, etc.

Table 5-4

Order lead time, days	Stockout cost per stockout per day	Expected number of stockout days per order	Stockout cost per order	Stockout cost per year
16	$.50	57.75	$28.88	$115.50
18	.50	27.95	13.98	55.90
20	.50	10.43	5.22	20.86
22	.50	2.57	1.29	5.14
24	.50	0.30	0.15	0.60
26	.50	0	.00	.00

Table 5-5

Order lead time, days	Holding cost per year (Table 5-3)	Stockout cost per year (Table 5-4)	Total cost per year		Total cost per year if cost per stockout only 5¢ per day
16	$ .00	$115.50	$115.50		$11.55
18	0.28	55.90	56.18		5.87
20	2.16	20.86	23.02		4.25
22	7.08	5.14	12.22		7.59
24	15.04	0.60	15.64		15.10
26	24.60	.00	24.60		24.60

be $3.3 \times \sum_{i=1}^{n} i$. (If 5 days late, for example, $\sum_{i=1}^{5} i$ would be $5 + 4 + 3 + 2 + 1$, or 15.) If the cost of a stockout day is assumed to be 10 percent of the item cost, or 50 cents, a stockout cost table similar to the holding cost table (Table 5-3) can be constructed by the above method of calculation. Table 5-4 is such a table.

The optimum order lead time is that which minimizes the total cost of holding plus stockouts. Table 5-5 computes the total cost for each lead time, and finds that it is lowest at a lead time of 22 days. In order to test the sensitivity of this finding to the most uncertain figure—the estimate of unit stockout cost—the total is recomputed on the basis of a cost per stockout day only a tenth as large as that assumed above. The results of this, in the right column of Table 5-5, show an optimum lead time of 20 days—not very different from the previous optimum.

A decision to "play it safe" on inventories by ensuring against any stockouts is seldom the optimum decision. The only way to be sure is to assess the unit costs as carefully as possible and make the comparison analytically.

DEMAND ANALYSIS

Purpose

Up to now the problems have been those on the supply side. Usage rate has been assumed known, and you have worried about how to stock material in the best way to meet this rate.

This is a reversal of real business priorities. The problem of inventory timing must be solved, to avoid unnecessary costs, but usually determining how much inventory you need in the first place is more important. This section will look into the problems of selecting an optimum inventory policy when demand is uncertain.

Expected value

Expected Value of a Gamble

If you toss three dice (a red one, a white one, and a green one), what is the *likelihood* of getting exactly 1 six? There are three ways to get such an outcome:

	Red die	White die	Green die
First way	A six	Not a six	Not a six
Second way	Not a six	A six	Not a six
Third way	Not a six	Not a six	A six

The probability of the first outcome—an event which has a 1/6 chance of occurring followed by two events, each of which independently has a 5/6 chance of occurring—is 1/6 × 5/6 × 5/6, or 25/216. The probability of the second outcome is the same, as is the probability of the third outcome. The probability that you will get one of these three outcomes is the sum of these three separate probabilities: 75/216, or a bit over 1 chance in 3.

Now for a different question about probability. If you toss three dice, what is the *expected number* of sixes you will get? (This is another way of asking what the *average number* of sixes per throw would be, if you tossed the three dice thousands of times.) Perhaps you see intuitively that, since the expected number of sixes per throw of a *single* die is 1:6 (in the long run, that is, you would expect a six about once every sixth throw), the expected number of sixes per throw of three dice is 3 times this much, or 1:2.

You can systematize this intuitive conclusion with a table of the following form, which will be useful in problems too complex for intuition:

Event	Probability of event	Conditional value (payoff if event* should occur)	Expected value (payoff times its probability)
Six on red die	1/6	1	1/6
Six on white die	1/6	1	1/6
Six on green die	1/6	1	1/6
		Total expected value	1/2

* The 1 means a six is thrown once.

To put this in terms of a gamble, suppose a casino charges you $1 to make such a toss, but pays you $2.50 for every six that comes up.

PROFITABLE INVENTORY MANAGEMENT

What is the expected value to you of this proposition? Make another table like the one above:

The event	Probability of event	Conditional value (payoff if event should occur)	Expected value (payoff times its probability)
No payoff	1/2	−$1.00	−$0.50
Payoff	1/2*	+$2.50 − $1.00	+ 0.75
		Total expected value from a toss	+$0.25

* Note that this is not the probability of getting exactly 1 six. That was computed earlier to be 75/216. It is the average number of sixes you will get per throw. Out of 216 throws, in the long run, 125 will be 0 sixes; 75 will be 1 six (paying off $2.50); 15 will be 2 sixes (paying off $5.00); and 1 will be 3 sixes (paying off $7.50); the total payoff is 108 × $2.50, equivalent to a payoff every other throw.

Utility of a Gamble

One point should be made about expected-value analyses. They are in monetary terms, and they ignore the fact that you assess the value of a dollar differently under different circumstances. If the casino offered you a contract to play the above game 100,000 times, no doubt you'd accept eagerly: but if it offered to play the game with you once, with the stakes multiplied by 100,000, it would be a different ball game. The monetary expected values are the same, but the utilities to you are not; when the possible loss approaches the limit of your resources, you cease being a "rational" man.

In the above case, a high possible one-time loss caused you to reject a gamble with positive expected value. Similarly, a high possible one-time reward can cause you to accept a gamble with a negative expected value. If a $2 sweepstakes ticket gives you 1 chance in 100,000 of winning $50,000, the expected value would be as follows:

Event	Probability	Conditional value	Expected value
Don't win	99,999:100,000	−$2	−$2.00 (almost)
Win	1:100,000	+$49,998	+ .50 (almost)
		Total expected value of ticket	−$1.50

The expected value is negative, but $50,000 is so high relative to your resources that its positive utility even at such hopeless odds overbalances the negative utility of a loss of $2.

Such actions aren't irrational, for the effect of resource limitations must be taken into account, but they do depart from mathematical rationality. In the remainder of this chapter it will be assumed that your level of financial operation does not approach your resource limits closely enough to cause such departure.

Expected Value of Stock

Many texts turn to the newsboy for help at this point, as an independent merchant who doesn't clutter his accounts with overhead costs. Suppose he buys 100 papers a day at 4 cents each, and sells all he can at 10 cents each. Past experience tells him that, while market demand for any specific day is unpredictable, its pattern can be predicted as follows:*

20% of the time there will be demand for exactly 25
20% of the time there will be demand for exactly 50
40% of the time there will be demand for exactly 75
20% of the time there will be demand for exactly 100

The expected value of stocking 100 papers can be calculated in the same way as the expected value of a gamble—indeed, the paper boy's decision to stock 100 papers *is* a gamble. If he sells 25 papers, he spends $4 and takes in only $2.50; the conditional value of a demand of 25 is therefore $-\$1.50$, and so forth.

Event	Probability	Conditional value	Expected value
Demand for 25	0.2	−$1.50	−$0.30
Demand for 50	0.2	+ 1.00	+ 0.20
Demand for 75	0.4	+ 3.50	+ 1.40
Demand for 100	0.2	+ 6.00	+ 1.20
		Total expected value of stocking 100	$2.50 per day

Suppose he selected a different stock level, say 75. The calculation is the same but the conditional values are different:

Event	Probability	Conditional value	Expected value
Demand for 25	0.2	−$0.50	−$0.10
Demand for 50	0.2	+ 2.00	+ 0.40
Demand for 75	0.4	+ 4.50	+ 1.80
Demand for 100†	0.2	+ 4.50	+ 0.90
		Total expected value of stocking 75	$3.00

* This highly artificial distribution (there is never a day when 33 are demanded' for example) is selected to make the example simpler. The technique works just as nicely with a more realistic distribution.

† But only 75 are sold, since this is all the stock he has. The above expected value calculation includes no "cost" for disappointing customers.

PROFITABLE INVENTORY MANAGEMENT

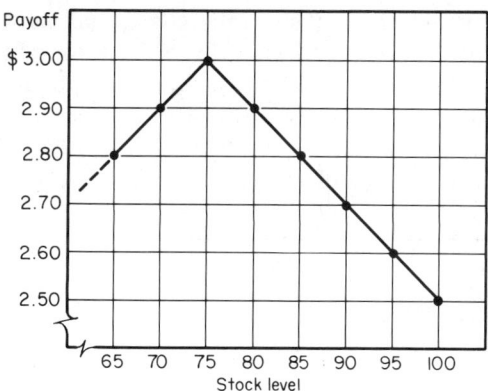

Fig. 5-7 Stock level vs. payoff.

He could calculate the expected values of all possible stock levels from 0 to 100, and present them in a "payoff table" similar to Table 5-6.

Figure 5-7, a payoff curve for this part of the payoff table, shows the pattern of change in expected profit with stock level. For an amount stocked less then 75, every increase of 1 paper in stock level increases expected profit 2 cents a day. At the optimum point this changes abruptly, and thereafter every increase of 1 paper in stock level *decreases* expected profit 2 cents a day. It should be possible to calculate the point at which this increment of additional profit first becomes negative— and a stock level 1 less than this would be the optimum strategy point. This calculation will be undertaken in the next section.

Incremental analysis

Constant-profit-rate Regions

Figure 5-7 shows that there are at least two different profit regions, in each of which expected profit changes at a different rate with increase

Table 5-6 Payoff table for different stock levels of newspapers

Stock level selected		Expected profit per day
70		$2.90
74		2.98
75	Optimum act	3.00
76		2.98
80		2.90
85		2.80
90		2.70
95		2.60
100		2.50

in stocking level. In the region below 75, the profit change per unit increase in stock level is $+2$ cents, and in the region above 75, the profit change is -2 cents. There are two more regions it will pay to explore: the region below a 25-paper stock level, and the region between a stock level of 25 and 50 papers.

First, restate the demand distribution of this example:

No. of papers demanded	Probability that demand is this number	Cumulative probability that demand is this much or more
25	0.2	1.0
50	0.2	0.8
75	0.4	0.6
100	0.2	0.2

For the analysis, let:

C = cost in dollars
R = receipts in dollars
P = profit in dollars ($P = R - C$)
S = daily stock level

Stock level between 0 and 25 (region I)

Costs: $C = \$.04 \times S$
Receipts: Since you are sure, in this region, of selling all you can stock (cumulative probability, above, equals 1), $R = \$.10 \times S$
Profits: $P = R - C = \$.10 \times S - \$.04 \times S = \$.06 \times S$
(Your profits are rising 6 cents for each additional paper you stock. So far, so profitable; now look at the next region.)

Stock level between 25 and 50 (region II)

Costs: $C = \$.04 \times S$
Receipts: In this region, you are sure of selling the first 25 papers, but only 80 percent sure of selling a twenty-sixth, a twenty-seventh, etc., up to a fiftieth, as your cumulative probability table tells you.

R = certain receipts from sales of first 25 papers + 80% probability of receipts from sales after twenty-fifth paper
$R = (\$.10 \times 25) + (\$.10 \times (S - 25) \times 0.8)$
$R = \$.50 + \$.08 \times S$

Profits: $P = R - C = \$.50 + \$.04 \times S$
(Your profits are rising 4 cents for each additional paper you stock, as the term $\$.04 \times S$ tells you. Keep calculating!)

PROFITABLE INVENTORY MANAGEMENT

Stock level between 50 and 75 (region III)

Costs: $C = \$.04 \times S$ (as always)

Receipts: In this region you are only 60 percent sure of selling a fifty-first, a fifty-second, etc., up to a seventy-fifth paper.

$R =$ receipts from sales of first 50, from formula above $+$ 60% probability of receipts from sales after fiftieth paper

$R = (\$.50 + \$.08 \times 50) + (\$.10 \times (S - 50) \times 0.6)$

$R = \$1.50 + \$.06 \times S$

Profits: $P = R - C = \$1.50 + \$.02 \times S$

(Your profits are rising 2 cents for each additional paper you stock, as the term $\$.02 \times S$ tells you; this confirms your previous findings. Keep calculating.)

Stock level between 75 and 100 (region IV)

Without going through the calculations—which are similar to those above—the profit equation will be $P = \$4.50 - \$.02 \times S$.
(Your profits are starting to fall, telling you not to stock the seventy-sixth paper. This does not mean that stocking 76 papers will give you a loss; but only that it will give you a profit 2 cents less than stocking 75 papers.)
The profit curve for these four regions is shown in Fig. 5-8.

Marginal profit analysis

It would be nice to avoid this step-by-step process and calculate the optimum stock level directly. The marginal profit method* makes it possible to do this.

* Developed by Robert Schlaifer, and described in detail in his book *Probability and Statistics for Business Decisions*, McGraw-Hill Book Company, New York, 1959.

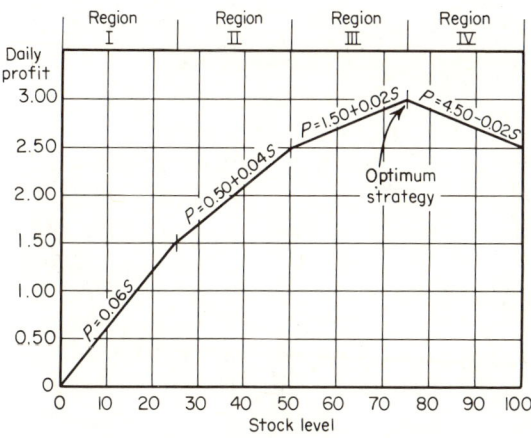

Fig. 5-8 Daily profit vs. stock level.

Instead of dealing with the expected profit of any given stock level, the marginal analysis method calculates the expected profit you will get from "just one more item" and finds the point at which this first becomes negative. You already know that this will occur at the seventy-sixth paper stocked, but now you want to determine it analytically.

Imagine that you are creeping up on the answer by trying the previous calculations with a stock level of 1, 2, 3 and so on; or in general, a stock level of n papers a day. A *payoff table* for stocking this nth paper, which will be true for any value of n, would be as follows:

	Stocking action	
Event	Stock	Do not stock
No demand for nth unit	−$0.04	$0
Demand for nth unit	+$0.06	$0

An *expected-profit table* for stocking the nth unit where the unit cost of a paper that is unsold is designated $c(=4$ cents$)$ and the unit profit of a paper that is sold is designated $p(=6$ cents$)$, would be as follows:

Event	Probability	Conditional profit	Expected profit
Demand less than n	$P(D < n)$	$-c\ (= -0.04)$	$-c \times P(D < n)$
Demand equals n or more	$P(D \geq n)$	$+p\ (= +0.06)$	$+p \times P(D \geq n)$
		Expected profit of stocking nth unit (adding)	$p \times P(D \geq n) - c \times P(D < n)$

You are trying to find out what your profit will be if you stock "one more item" at any level. The *first column* of the table lists the two possibilities: the number demanded is less than n (you will not sell the nth) or the number demanded is n or more (you will sell the nth). The *second column* simply expresses these two probabilities in mathematical notation. The *third column* lists the profit you will make if you don't sell it (you lose 4 cents) or if you do sell it (you make 6 cents). The *fourth column* multiplies each of these two conditional profits by its probability of occurrence, in the same way you have done with the previous expected-value tables. The sum of the fourth-column values is the expected profit from stocking the nth unit.

How do you find the probabilities in the second column? Refer to the table on page 86: the probability that demand is equal to or

PROFITABLE INVENTORY MANAGEMENT 89

greater than some value of n, say $n = 50$, is shown from that table to be 0.8—and the probability that demand is less than 50 must be 0.2. [For $n = 51$, $P(D \geq 51)$ is 0.6, and $P(D < 51)$ is 0.4.]

Look at the sum of the values in the fourth column of page 88—the expected profit from stocking the nth paper. Profit for stocking the nth unit will be positive—that is, it will pay the newsboy to add that nth unit to his daily stock rather than stocking $n - 1$ newspapers a day—as long as the positive part of this sum is greater than the negative side of it: as long as $[p \times P(D \geq n)] > [c \times p(D < n)]$.

If you substitute in this formula $P(D \geq n) = 1 - P(D < n)$ (which must be true since the two probabilities have to total to certainty, or 1), it reduces to the following decision rule:

$$P(D < n) < \frac{p}{p + c}$$

This means, in words,

> Continue increasing stock level as long as the probability that demand is less than n papers is less than the ratio: profit/(profit + cost).

Try this formula on $n = 75$, which you know from Fig. 5-8 is the optimum strategy. The probability that demand is *less than* 75 papers (that is, 25 papers or 50 papers) is 0.4, and this value is less than $0.06/(0.06 + 0.04)$, which is 0.6; so you would continue increasing stock level to 75. Try 76 papers: the probability that demand is *less than* 76 papers (that is, 25, 50, or 75 papers) is 0.8, and since this is *not less than* the ratio you would not stock a seventy-sixth paper. The formula leads you directly to the optimum value, if you know the cumulative demand probabilities and your unit costs and profits.

The value of information

A newsboy who goes through this process is pretty smart, and certainly merits the title "rational manager." But he still pays a price of some uncertainty: he knows the *distribution* of expected demand, but not its precise outcome. That is, he knows that over the long run:

> 25 papers will be demanded 2 days out of 10
> 50 papers will be demanded 2 days out of 10
> 75 papers will be demanded 4 days out of 10
> 100 papers will be demanded 2 days out of 10

But since he does not know which of these days is which, the best he can do is play the odds and calculate that a stock of 75 is his optimum

Table 5-7

Demand	$P(D)$	Actual paper boy			Superhuman paper boy			Paper boy's losses from incorrect stocking	
		Stock	Conditional (P)	Expected (P)	Stock	Conditional (P)	Expected (P)	Over	Under
25	0.2	75	−$ .50	−$ .10	25	+1.50	+$ .30	$.40	
50	0.2	75	+ 2.00	+ .40	50	+3.00	+ .60	.20	
75	0.4	75	+ 4.50	+ 1.80	75	+4.50	+ 1.80		
100	0.2	75	+ 4.50	+ 0.90	100	+6.00	+ 1.20		$.30
			EXP$(P)_{act}$ = $3.00			EXP$(P)_{sup}$ = $3.90			

Overstocking loss $.60
Understocking loss $.30

Estimated value of perfect information $.90

choice in the face of this uncertainty. This is a perfect choice on the days that demand is 75, but it is less than perfect on the other days.

Compare his performance with that of an even brainier paper boy, who can predict the days for each demand and stocks accordingly. This comparison, Table 5-7, shows that the first paper boy, while doing the best that could be expected of him, could make an estimated 90 cents more a day if his information were "perfect."*

This daily sum is his best present estimate of the maximum possible profit improvement for complete information. Its usefulness lies in the fact that it is a firm ceiling on the amount he ought to spend (in market surveys and so on) in an effort to improve his information about demand. Since he will not hope for 100 percent success, the practical ceiling is a good deal less than this.

SUMMARY

This chapter dealt first with the inventory problems involving the supply side, assuming for the moment that demand was constant, or at least was known precisely. Then it went into the analysis of demand, and

* Note that he might not *really* make 90 cents more a day—perhaps in the future 100 papers will be demanded every single day; this is his *present estimate* of the additional profit potential if he could call each day's demand. As in all managerial decision-making, you never actually see the future; you do your best on the basis of your systematic *expectations* about the future.

developed the concept of maximizing expected value in the face of probabilistic demand.

In any real problem the manager will face such problems on both the supply and the demand side simultaneously. This chapter assumed, for clarity, that the problem was on the supply or the demand side only, and not on both sides at the same time; but the manager seldom will be so fortunate.

6

Capital Expenditures

INTRODUCTION

Definition of the problem

Governments, businesses, and individuals buy capital goods to bring maximum satisfaction per dollar, or at least they try. Individuals make many subjective judgments of the pleasure that prospective purchases will bring; but in items such as cars and power mowers they often are guided by quantitative figures on performance and service life—and the popularity of consumer research journals shows that such precise information is welcome. The federal government is spending a sizable bankroll to polish its "benefit-cost" techniques for assessing all the benefits of competing public projects, and some of the methods display a surprising sophistication. One might think that the existence of profit as an overriding criterion of choice would give business firms all the best of it, but as businessmen know, it isn't quite that simple.

The capital investment problem can be stated easily: how to allocate a given level of resources between alternative opportunities to acquire fixed assets, in a way that will maximize overall profits. There are basic choices to be made: to buy or rent plant or tools, to expand present facilities or not, to automate or be labor-intensive, to repair or replace equipment, to modernize or stay less efficient, to diversify lines or remain selective. Once these are resolved, there are more detailed choices: to buy X or Y, to expand at home or in a new location, how much to automate, how frequently to repair or replace, and so on.

Nothing about these problems, as stated, is unduly complicated. If you have clear-cut economic choices, and know their costs and benefits, the accounting techniques for choosing between them are logical and well understood. But there are elements of uncertainty.

Uncertainty about alternatives. When you face two alternatives—to replace facilities or to continue using what you have—other investment opportunities are lurking in the wings. Never is the choice confined to present alternatives, because tomorrow a better selection might show up and without liquidity you cannot seize it.

Uncertainty about numbers. If you don't know what your proposed capital projects will earn, or their productive life, or their true cost, you can't use those logical analysis techniques.

Uncertainty about your firm's viewpoint. Most firms do not express their objectives and attitude toward risk numerically; yet without such numerical expression, there is no systematic way to choose between, say, a $1 million investment with some specified probability of being profitable and a $10 million investment with some specified better probability of being profitable (but of being a whopper if it fails).

Uncertainty about how fixed is "fixed." Capital expenditures are termed "fixed," but this term is subject to infinite variations. Some purchases commit you to a specific use for many years, while others can be adapted to a variety of uses. There are ticklish choice problems involved in determining how much more to pay for a capital item that is not completely fixed, but is flexible in its employment.

Strategy versus tactics

Capital investment problems encompass virtually the total spectrum of management decisions. The broadest question of major policy, such as "Shall we go international?" or "Shall we stop making toasters and enter the race for space contracts?" are problems of capital investment; and the costs of girding to enter a new competitive field can far exceed the cost of buildings and equipment. But these momentous affairs, corresponding to strategic military decisions made at the center of government, are beyond the scope of this chapter.* The discussion that follows will deal with tactical decisions—how to make optimum choices when the general path is determined and the question is one of selecting between alternate means.

* For an interesting treatment of capital investment questions in the large, see H. I. Ansoff, *Corporate Strategy*, McGraw-Hill Book Company, New York, 1965.

Types of capital goods

Of many classifications that might be used, one that makes sense from an operations research viewpoint differentiates on the basis of how the capital item loses utility:
1. Those that deteriorate in performance over time (and may be restored in whole or part by expending maintenance funds)
2. Those that perform adequately until sudden complete failure (the length of service until failure varying more or less randomly over some predictable range)

Buildings of course fall within the first class, as generally do complete items of equipment. Light bulbs fall within the second class, as generally do individual components or repair parts for equipment. But a vital component in the latter class can transform the whole equipment of which it is a part into that class *for a particular mode of operation* (as your wife finds when she has a flat tire on the turnpike, or the production manager learns when a control device failure idles the production line). So if nonstop performance is worth money, you may classify the capital item differently than if maintenance shutdowns can be accepted. You must decide which class you are dealing with, because the arithmetic used to model the behavior of the two classes is a bit different.

PROFITABILITY ANALYSIS

Single-product case

Costs

It isn't very profound to say that costs can be classified into fixed costs that don't change with quantity sold, and variable costs that do—nor is it really true. But it is a good concept as a start.

Figure 6-1 shows these simple relationships. At zero volume, all costs are those basic to being in business at all—rent, basic utilities, flat salaries, depreciation, and so on. As volume picks up, the basic costs change little, but you start to incur a new set of costs that grow with production—materials, variable utilities, hourly wages of production employees, shipping, and so on. Since the basic costs don't rise, your unit costs fall continuously as volume grows within normal plant capacity. If you put some numbers on the graph of Fig. 6-1—say a fixed cost of $500 and a variable cost that increases $50 per unit—this relationship is shown by the unit cost curve of Fig. 6-2.

CAPITAL EXPENDITURES

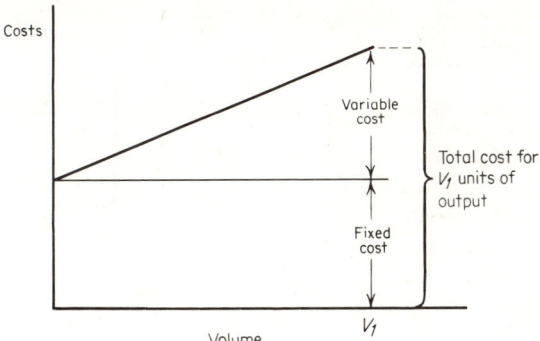

Fig. 6-1

    This simple curve makes some unrealistic assumptions (aside from the question of whether you really can assign fixed costs to products). It assumes that fixed costs are unaffected by volume; which isn't completely true even in the very short run because use of facilities increases the tempo of basic functions somewhat, and isn't true at all for longer periods because production ultimately demands overhead expenditures which would not be incurred if you shut down. It assumes that variable costs increase equally for each unit of output, whereas a more typical curve might start out steeply when you are having growing pains, level off in the middle as you hit your stride, and rise again at the end as you bump into full-capacity bottlenecks. It ignores the difference in slope between constant-volume production, where you engineer for maximum economy, and ever-shifting output, where you keep incurring sizable get-ready costs. A more realistic curve would look like Fig. 6-3.

    The situation is not, however, as bothersome as it looks. For a given mode of operation (steady run or varying runs), and within efficient operating bounds for your business, you can abstract from Fig. 6-3 a nearly linear part where your greatest interest lies—and it does fit the simple model passably well, as Fig. 6-4 shows.

Revenue

Economists say that, while an industry selling price must fall as volume goes up and the market approaches saturation, a single producer's selling price remains constant as his volume rises, since his operations are too small to affect the industry market. Every corner grocer knows this isn't true. The national market for X brand of coffee is huge, at $1.39 a jar. Maw-and-Paw's Grocery can sell a certain number of jars at $1.39 or a bit more, by cutting the price to $1.19 it can sell a few extra to the regular customers, and by slashing to 59¢ it can steal business from Jake's Delicatessen and some other stores. But there is some volume above

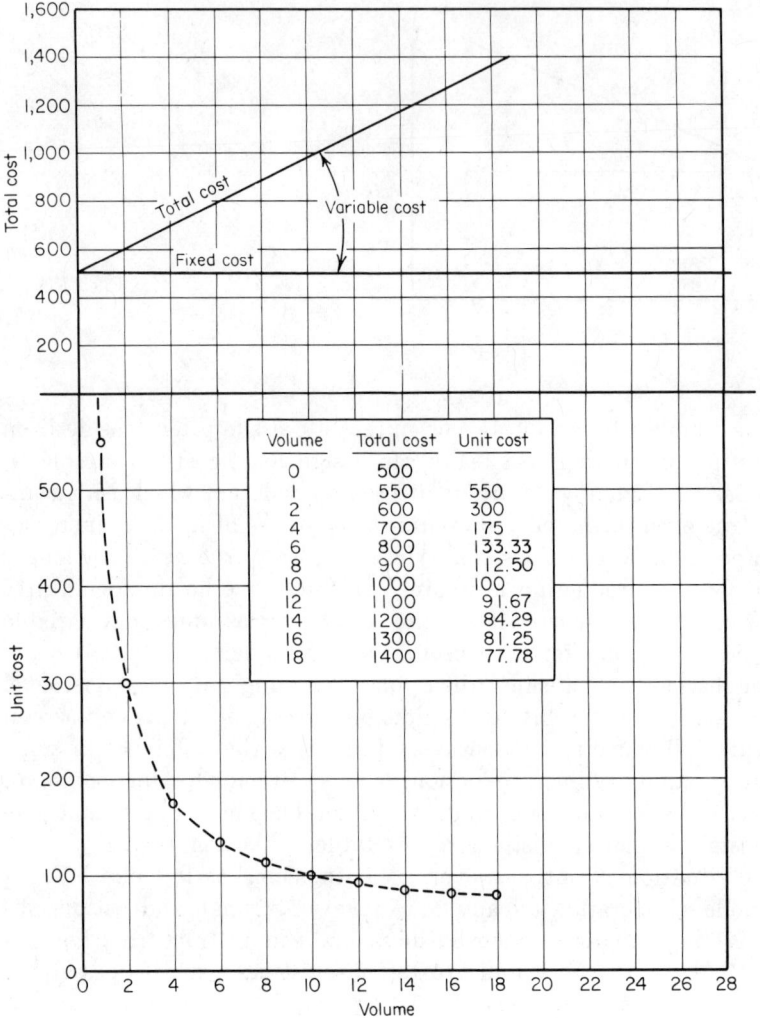

Fig. 6-2

which it'll scarcely be able to give X brand away. The industry market hasn't been saturated, but Maw-and-Paw's market has.

Every business has a certain base volume arising from the fundamental characteristics of location and past performance, and it can be increased only by accepting less net revenue per unit of output.* The decrease may come from price reduction or from marketing efforts which are incurred to purchase more and more volume at less and less incremental gain.

* As discussed in Chap. 3.

CAPITAL EXPENDITURES

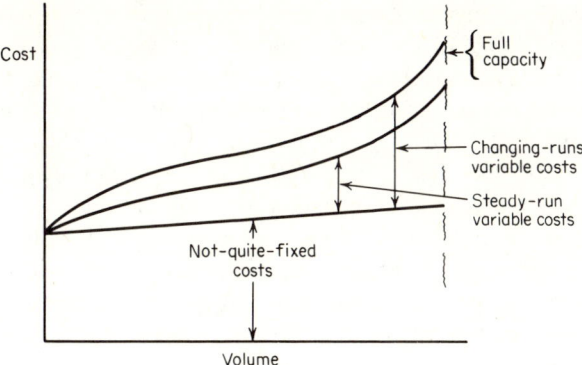

Fig. 6-3

Net profit is not going down—the principal reason for engaging in such strategy is to push net profit up—but each new increment of business brings less *unit* profit than the last.

Break-even Analysis

These two types of costs—production costs, for getting units of output ready for sale, and marketing costs, for getting customers to buy them—are quite different in concept. If you produced 1,000 widgets in one month, the production cost would be the same whether you sold them that month or not; but the selling costs might be quite different according to whether you tried to sell them all in a month or took six months to sell them. Since the cost of marketing simply reflects the phenomenon of diminishing returns, conceiving of such a cost as reduction in revenue makes for clearer profitability analysis.

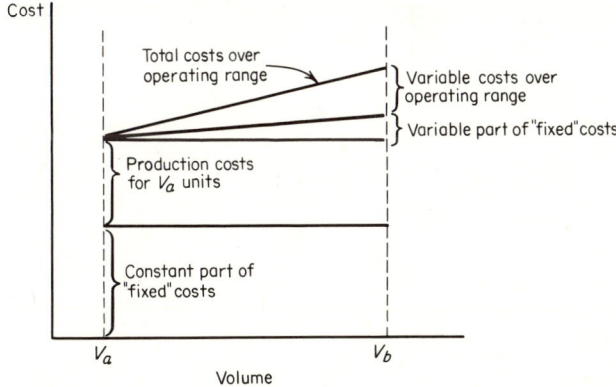

Fig. 6-4

Figure 6-5 superimposes curves of cost, and revenue (after subtracting marketing cost), plotted against output volume. Below output V_0, where the curves first cross, the business operates at a loss. Above output V_1, where the cost curve (increasing as you near capacity) crosses the revenue curve (decreasing under the burden of more and more promotion), the business again operates at a loss. Output V_{opt} is the output level of maximum net profit. Some managers consider that operation at any other point incurs an "opportunity loss" of an amount by which it fails to match this maximum profit. Another viewpoint considers unit cost at the most efficient volume (generally somewhere near full production) as standard cost, and computes the higher unit cost at some lesser volume as a cost of nonproduction, so that the price paid and profit foregone by less than optimum volume can be recognized clearly.

Multiproduct analysis

The above analysis applies to a one-product firm. A real firm has many products, and the more of its resources it invests in one, the less it can invest in the others. If you have two products, X and Y, and their cost curves (including interest on the invested capital) are as shown on Fig. 6-6, you can determine the least-cost distribution of your resources between the two products by trial and error. A more systematic way is to plot the cost curve for Y from the other end of the scale, total the two separate cost curves at each point, and find the optimum mix from the lowest point of the total-cost curve. In the illustration, this mix would be 46 percent to product X and 54 percent to product Y.

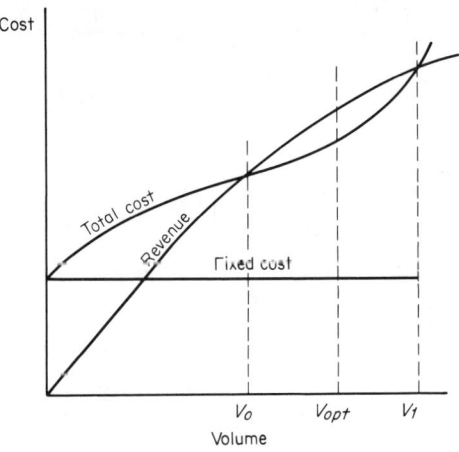

Fig. 6-5

CAPITAL EXPENDITURES

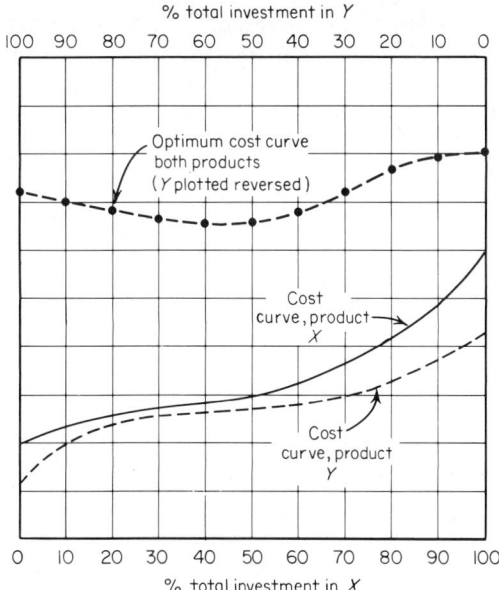

Fig. 6-6

A similar approach can be adopted for determining the incremental rate of return for two projects.*

The problem of maximizing the return from more than one product is quite similar in concept to the problem of capital investment—choosing between several capital investment alternatives in a way that will maximize return. The problem discussed above involved selecting optimum proportions of two alternatives; when more than two are involved, the problem grows rapidly in complexity, and requires the methods of mathematical programming described in Chap. 10 for its solution.

CAPITAL EXPENDITURE ANALYSIS

A checklist for appraising capital projects

It is convenient to approach the analysis of capital investments by looking at the problem from five viewpoints, each invoking a different sort of managerial expertise.

---

* For a full discussion of the allocation of funds to competing projects, see James C. Hetrick, Mathematical Models in Capital Budgeting, *Harvard Business Review*, January-February, 1961, p. 49.

1. *Production factors:* costs of labor and material, demand on other capital stock, distribution and handling costs, maintenance costs, etc.
2. *Marketing factors:* costs of promotion, estimated sales, strength of competition, distribution of sales over time, etc.
3. *Accounting factors:* annual income, project life, depreciation rate, investment requirements
4. *Risk factors:* uncertainty of cost and revenue estimates, equipment reliability, etc.
5. *Competitive investment opportunities:* other projects, known and unknown, which are alternative candidates for investment

Evaluation methods

Most analytical methods for evaluating the worth of projected capital investments involve a variant of one or more of the following techniques: (1) payback period, (2) return on investment, (3) present worth.

Payback Period

The payback period method measures how many years the capital investment will take to amortize itself through profits attributable to the investment. A major petrochemical firm, whose new products enjoy a relatively short earning life in this swiftly moving field, requires that a new proposal pay back in four years; industries with slower technology find this far too strict a requirement.

This criterion has several drawbacks. It tells nothing about the earnings trend of the investment at the end of its payback period—is the trend almost played out or just beginning to boom? This method doesn't differentiate between investments on the basis of how fast they pay back within the period: each of two projects may return its investment cost in 10 years, but one may pay back the major share in the first year or so, while the other may be a slow bloomer that doesn't start earning until the last few years; and at 10 percent interest, a sum deliverable next year is worth nearly twice as much as the same sum deliverable in 8 years. This criterion makes no explicit allowance for differences in the riskiness of projects. Your judgment can adjust for these factors, but when it does you aren't really using the payback period criterion—you are using a more complex criterion whose rules are not stated specifically.

Return on Investment

This yardstick measures earnings as a fraction of the amount invested. The "original-cost basis" states its yearly earnings as a fraction of original cost. The "average-cost basis," assuming that the initial investment

CAPITAL EXPENDITURES

is paid off in regular increments over its useful life, takes the average investment as half the original investment (neglecting salvage value). The "discounted-cash-flow basis" finds the equivalent discount rate which, if applied to each year's estimated earnings, would make the combined present worth of these earnings equal to the initial investment cost.

Consider an investment which costs $50,000 and is estimated to earn $9,800 a year for a 15-year life, so that its net earnings are $97,000, or $6,500 a year.*

Its earnings rate on an *original-cost* basis will be $6,500/$50,000 or 13 percent.

Its earnings rate on an *average-cost* basis will be $6,500/$25,000 or 26 percent.

The *discounted-cash-flow* method assumes that annual earnings regularly pay back part of the initial investment, and computes an equivalent rate based on each year's remaining balance in the same way that a home mortgage collects interest on only the unpaid balance. Ask your accountant what interest rate would be involved in a 15-year mortgage for $50,000 with a $9,800 annual payment, and a look in his cumulative discount table will tell him 18 percent—our discounted-cash-flow earnings rate for this example.†

Present Worth

The third kind of yardstick compares investment candidates by converting the estimated future earnings stream of each into its equivalent present value, at some earnings rate that you consider about right for your business or industry. A good analogy is your insurance man's "single-premium annuity"; you tell him how much you want per year and for how long, and he looks up the lump-sum premium you have to pay now. This is a tidy method for estimating the relative values of two or more projects—provided everything else is equal.

Treatment of uncertainty

Everything else is not equal, of course; some propositions are gold-plated, some are harebrained, and the rest lie in between. Seldom do you face

---

* ($9,800 × 15) − $50,000.

† In other words, prospective annual earnings of $9,800 for 15 years have a present worth of $50,000 when discounted at 18 percent, and a "rational" businessman who felt money was worth 18 percent to him would be hard put to choose between the cash and the annuity. For a more detailed look at the wonders of discounting and write-offs, see a book such as R. N. Anthony, *Management Accounting*, Richard D. Irwin, Inc., Homewood, Ill., 1960, or the Machinery and Allied Products Institute formulas as described by G. Terborgh in *Dynamic Equipment Policy*, McGraw-Hill Book Company, New York, 1949.

the alternatives of a rich strike or bankruptcy, but you may often face the choice between a fairly sure prospect with moderate return and a not-so-sure one which could bring substantially more—but might bring substantially less. The previous methods are helpful when you are in territory so familiar that your estimates are ironclad, but how do you deal systematically with the element of risk?

A popular technique is to increase the discount rate: charge the riskier project more for its funds, to hedge against the prospect that it may earn less than your estimate. This is a dubious method. A mechanical objection is that higher discount rates will lessen the present value of future *costs* as well as future *revenues*, and if the uncertainties about future costs and revenues are unequal, you may in fact be helping the uncertain project to be chosen when you intended to do the opposite.

This objection can be handled by a skillful accountant, who increases the discount rate for revenues and decreases it for costs. But a more fundamental objection is the distortion of your estimates, imposing an across-the-board conservatism you may not intend. It is worth looking at exactly what you do by such a technique.

You must choose, let us say, between two investments which will pay off in a lump sum 10 years from now, each requiring the same investment today. The first will return $35,000 with virtual certainty. The second has a 50 percent chance of returning $38,000, but there is a 20 percent chance of its returning $32,000 and a 30 percent chance of its returning $44,000. If your usual discount rate for capital projects is 15 percent, a 15 percent discount table tells you that the present value of a dollar received 10 years from now is 24.7 cents so your safe $35,000 project is worth $8,650. You decide that the uncertainty of the second project warrants making it carry a 20 percent discount rate,* for which the present value of a dollar 10 years hence is 16.5 cents. The present worth of your three earnings estimates then becomes

$5,160 with a 20 percent probability
$6,130 with a 50 percent probability   or a present "expected
$7,110 with a 30 percent probability   value" of $6,230†

Your board of directors, presented with this choice, must pick the first project—$8,650 is better than $6,230. (If the discount rates had been as close as 10 and 12 percent, the first project still would have won out, $13,500 to $12,400.)

This mechanized decision probably does not represent their real attitude toward risk. In the long run, since projects of the second type

---

* Higher than your usual 15 percent to compensate for the lesser probability of ultimate payoff.

† Obtained by multiplying each present value by its probability, and summing.

earn 10 percent more than the first, a company large enough to undertake many projects and spread the risk would not turn its back on higher earnings. When estimators have made careful estimates of the earnings spread, nothing is accomplished by depressing the earnings artificially with a higher discount rate. What you want is a realistic statement of the estimated earnings *together with the odds*. If your first project is discounted at 15 percent, your second project also ought to be discounted at 15 percent, with its probable earnings spread presented as accurately as possible so that a decision can be made on the basis of all the information. Then you could say about the second project: Its present worth is $7,900 with a 20 percent probability, $9,380 with a 50 percent probability, and $10,870 with a 30 percent probability—or an expected value of $9,530.* *This* is what the board should compare with the certain present worth of $8,650 for the first project.

How does the board reach its decisions in such cases? Certainly it has to know its own attitude toward risk, and has to have a good feeling for how large a gamble it can afford. Companies undertake speculative projects as a matter of course, when the cost of each is small relative to total worth (oil exploration is a good example), but think long and hard before hazarding funds that strain their resources to the limit.† John Von Neumann and Oskar Morgenstern were the first to suggest a concept for putting numbers on such feelings about risk, in their *Theory of Games and Economic Behavior*.‡ The art of reducing managerial hopes and fears to quantitative terms still is a bit too shaky for everyday use in problems of choice, but it is well to know where it is headed.

REPLACEMENT ANALYSIS

The preceding has considered a single capital investment or at most a mix of two, and has assumed that it will work for you faithfully throughout its stated life and then expire. This section will depart from such simple assumptions. At the start of the chapter a distinction was made between equipment that degrades in performance over time (unless maintained, and perhaps even then) and equipment whose usefulness ends by sudden

* Chapter 5 has looked more deeply into this matter of establishing the probabilities of various outcomes in management situations and of making effective use of such probability distributions in decision-making.

† The International Business Machine Company's staggering $5 billion investment in its 360 computer line (*Fortune*, September and October, 1966) is a rare type of corporate gamble.

‡ Princeton University Press, Princeton, N.J., 1947. For a less formidable presentation, see Billy E. Goetz, *Quantitative Methods: A Survey and Guide for Managers*, McGraw-Hill Book Company, New York, 1965.

failure on a more or less random basis. The remainder of the chapter will consider the management problems posed by these two classes of equipment, extending the previous analysis to more complex combinations of capital items.

Equipment that deteriorates

One Type of Equipment

Suppose you purchase a piece of equipment today, for P dollars, and anticipate that its maintenance cost will rise each year that it is in use: M_1 dollars the first year, M_2 the second, and so on. Its cost per year for 4 years would be

$$\frac{P + M_1 + M_2 + M_3 + M_4}{4}$$

Its cost per year for any number of years, say Y years, would be

$$\frac{P + M_1 + M_2 + \cdots + M_Y}{Y}$$

If you plan to keep this equipment Y years, then replace with a second just like it which you will keep Y years, then a third and so on, your cost for R such replacements can be expressed in a set of R expressions like the one above. To identify each equipment in the (rather lengthy) expression, you give it a subscript of 1 if it is the first, 2 if the second, and so on.

The cost per year for R such equipments, each lasting Y years, would be:

$$\frac{1}{YR} [(P_1 + M_{11} + M_{12} + \cdots + M_{1Y}) \\ + (P_2 + M_{21} + M_{22} + \cdots + M_{2Y}) + \cdots \\ + (P_R + M_{R1} + M_{R2} + \cdots + M_{RY})]$$

The management scientist, who is as lazy as the next fellow, writes this in the following shorthand:

$$\frac{1}{YR} \sum_{i=1}^{R} \left(P_i + \sum_{j=1}^{Y} M_{ij} \right)$$

Several Types of Equipment

Suppose you are choosing between four different pieces of equipment that you believe will produce the same output at the same operating cost, so that the choice depends on your estimates of their respective purchase prices, maintenance costs, and service lives. You have the following

CAPITAL EXPENDITURES

estimates of these elements:

Equipment No.	Purchase price, P	Annual maintenance cost over life service					Life, Y
		M_1	M_2	M_3	M_4	M_5	
1	$ 9,000	$500	$800				2 years
2	12,000	400	600	$1,200			3 years
3	15,000	300	500	1,300	$2,000		4 years
4	20,000	200	400	1,000	1,500	$1,900	5 years

The cost per year for each of these equipments is a simple calculation. The management scientist would say it:

$$\frac{1}{Y} \cdot \left(P + \sum_{j=1}^{Y} M_{ij} \right)$$

Cost per year for No. 1 = ½(9,000 + 500 + 400) = $5,150
Cost per year for No. 2 = ⅓(12,000 + 400 + 600 + 1,200)
  = $4,733
Cost per year for No. 3 = ¼(15,000 + 300 + 500
  + 1,300 + 2,000) = $4,750
Cost per year for No. 4 = ⅕(20,000 + 200 + 400
  + 1,000 + 1,500 + 1,900) = $5,000

If all your estimates are exactly correct, No. 2 (or No. 3) is the machine to buy.

Inclusion of Interest Rate

The compound interest formula gives the future value F of a present amount I if invested at a specified interest rate r for any given number of years (t):

$$F = I(1 + r)^t$$

Thus $100 invested at 10 percent for 3 years would be worth

$100(1 + 0.10)^3 = $100(1.331) = $133.10

Reversing the formula gives the present value of a future payment t years hence:

$$I = \frac{F}{(1 + r)^t}$$

Thus, the present value of $100 payable 3 years hence is

$$\frac{100}{(1.1)^3} \quad \text{or} \quad \$75.10$$

Now repeat the previous analysis, taking interest into account. To simplify the computations, assume that you make the entire year's maintenance payments at the end of the year. Machine No. 1, for example, would require a payment of $9,000 today, $500 a year from today, and $800 two years from today, at which time its useful life would be over.

Compare No. 1 with No. 3. The present values of the average yearly costs over the life of each machine are as follows:

$$\text{No. 1:} \quad \frac{1}{2}\left[9{,}000 + \frac{500}{1.1} + \frac{800}{(1.1)^2}\right] = \$5{,}058$$

$$\text{No. 3:} \quad \frac{1}{4}\left[15{,}000 + \frac{300}{1.1} + \frac{500}{(1.1)^2} + \frac{1300}{(1.1)^3} + \frac{2000}{(1.1)^4}\right] = \$4{,}510$$

But this isn't the whole story. At the end of the second year, No. 1 shuts down. A proper comparison with No. 3 requires you to buy another new No. 1 for the next two years. The four-year cost for No. 1 then becomes:

$$9{,}000 + \frac{500}{1.1} + \frac{800}{(1.1)^2} + \frac{9000}{(1.1)^2} + \frac{500}{(1.1)^3} + \frac{800}{(1.1)^4}$$
$$= 18{,}476 \text{ or } \$4{,}619 \text{ per year}$$

Proper discounting shows No. 1 to be a good deal more nearly competitive with No. 3 than the previous analysis indicated.

Optimum Replacement Time

The previous section decided which of several pieces of equipment with different lives would be cheaper, but did not consider the possible additional savings from modifying the replacement time. Suppose you are buying a machine for $1,000, with a 6 percent interest rate. You estimate that maintenance costs will be $60 at the end of the first year, $70 at the end of the second, and thereafter as shown in the second column of Table 6-1. The discount rate for the first year (present value of a dollar spent one year hence) is $1/(1.06)^1$ or 0.943; for the second year it is $1/(1.06)^2$ or 0.890, and thereafter as shown in the third column.

The fourth column (to continue filling in the table) is the present value of each year's maintenance cost, computed by applying the discount rate for a year to the maintenance cost in that year (thus, the present value of $120 spent for maintenance after four years of service is $95).

CAPITAL EXPENDITURES

Table 6-1

(1) Year t	(2) Maintenance cost	(3) Discount rate	(4) Present value of maintenance cost in tth year	(5) Present value of costs through tth year	(6) Cumulative discount rate	(7) Column (5) divided by column (6)
0		1	0	$1,000	1	$1,000
1	60	0.943	$ 56.6	1,057	1.943	543
2	70	0.890	62.3	1,119	2.833	395
3	90	0.840	75.6	1,195	3.673	325
4	120	0.792	95	1,290	4.465	285
5	150	0.747	112	1,402	5.212	269
6	180	0.705	127	1,529	5.917	259
7	210	0.665	139	1,668	6.582	254
8	240	0.627	150	1,818	7.209	252
9	300	0.592	177	1,995	7.801	256
10	400	0.558	223	2,218	8.359	265

The fifth column is the present value of the $1,000 purchase price plus the cumulative maintenance costs to the end of each year (thus the entry for year 3 is $1,000 plus the three entries down to that point in the fourth column). The sixth column is the cumulative sum of discount rates down to that point (the entry for year 1 is 1 + 0.943, for year 2 it is an additional 0.890, and so on).

The seventh column is the fifth divided by the sixth.

When you reach the end of any year—say the eighth—you can do either of two things:
1. Keep the machine a ninth year, then pay $300 maintenance and replace with another machine *that you keep for 9 years*.
2. Replace the machine now with another machine *that you keep for 8 years*.

Since it will take nine cycles of 8-year replacements and eight cycles of 9-year replacements before you first come to a year in which both machines will be due for replacement, the direct way to find which is the cheaper policy is to price out these two 72-year streams at their present value and compare them—quite a task. And you'd still have to investigate 7-year replacements, 10-year, and so on.

This table does the same thing for you. Compare column (2), which is constantly increasing, with column (7), which decreases and then increases. Replace the machine in the latest year that column (7) exceeds column (2). In the example, year 8 is the last such year [because in year 9 the maintenance cost of $300 in column (2) exceeds the value of 256 in column (7)]; therefore, you replace at the end of the eighth year.

Mathematical Models for Equipment Replacement

In the examples discussed thus far, it has been assumed that maintenance has been sufficient to keep the equipment performing at its original pitch until the end of its life. In the real world, obsolescence often reduces the utility of well-maintained equipment. Moreover, maintenance costs simply have been stated, with no effort to discern their underlying pattern. If there will be many such decisions, it is well to hunt for the mathematical expressions that depict reasonably well the drop in value and the rise in maintenance over time, so that you can take full advantage of the tremendous time saving offered by mathematical analysis.

How do you make mathematical models of this sort? You graph actual past data from your company, showing how value has fallen and maintenance has risen over time; your analyst finds mathematical expressions for curves with the same general shape as your graphs; you "fit" a suitable curve to each graph by adjusting the constant term or terms in your equations; and if the fit satisfies you (there are statistical tests for "goodness of fit," if you don't trust your eye) you set the data aside and use the formulas from then on.

The cost of a piece of equipment over a given number of years—say t years—has three elements:

$$\begin{matrix}\text{Purchase} \\ \text{price}\end{matrix} - \begin{matrix}\text{value remaining} \\ \text{after } t \text{ years}\end{matrix} + \begin{matrix}\text{maintenance} \\ \text{for } t \text{ years}\end{matrix}$$

and the cost per year for this number of years is the above expression divided by t. If you can express value and maintenance in formulas, the mathematical model builder uses a simple calculus operation to solve his equation for the value of t that gives minimum yearly cost.*

What do some of these mathematical models from your high school algebra look like? If you decide that both the decline in value and the total maintenance investment over the years are straight-line affairs that resemble Fig. 6-7, your equations are:

Value at year t: $\quad V_t = P\left(1 - \dfrac{t}{t_0}\right)$

Maintenance cost for t years: $\quad M_t = kt$

Put these into the cost-per-year expression above (the purchase price cancels out), and you get a cost-per-year value for any year, t:

$$(\text{Cost per year})_t = \frac{1}{t}\left[P\left(\frac{t}{t_0}\right) + kt\right]$$

---

* He sets the first derivative equal to zero and solves for t, in a process which any elementary calculus text will be happy to describe. You needn't know how to do it, but you ought to know that it can be done—and that it works.

CAPITAL EXPENDITURES

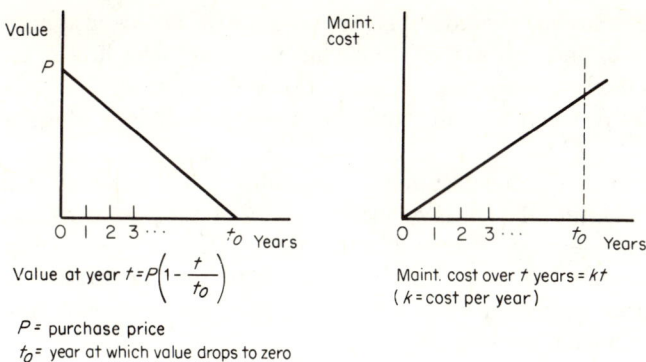

Fig. 6-7

P = purchase price
t_0 = year at which value drops to zero

When your industrial engineer uses his calculus to seek an optimum value of t, he finds there isn't any. The year of replacement has no effect on cost, *provided* your models are correct representations of reality.

Maybe they aren't. Maybe value or maintenance or both fit another curve—the exponential. (It wouldn't be surprising if they did; exponentials have proved fairly good for this purpose.)

If value drops exponentially, and maintenance is linear as before, your industrial engineer will find the best strategy one of keeping the machine as long as possible—the average cost per year keeps dropping. If both value and maintenance cost are exponential, as in Fig. 6-8, there is a specific minimum somewhere, and the simple calculus operation will find it.*

Equipment that fails

The other important class of equipment, of which the fluorescent lighting tube in your ceiling is typical, requires no maintenance but fails without

* For a more detailed look at this method, see Arnold Kaufmann, *Methods and Models of Operations Research*, Prentice-Hall, Inc., Englewood Cliffs, N.J., 1963.

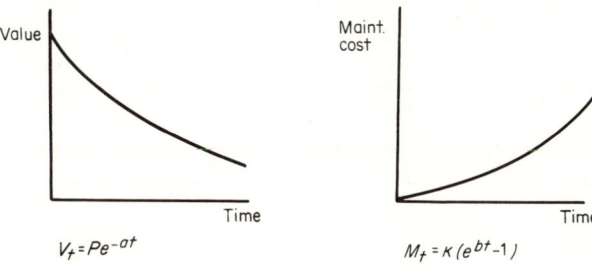

$V_t = Pe^{-at}$

$M_t = K(e^{bt} - 1)$

Fig. 6-8

warning. You are playing the odds here: you know *on the average* how long such tubes last, and records will tell you how much the life of an individual tube is likely to *vary above or below this average*, but within this range you can't know what an individual tube will do—it is a random process.

Fortunately, random processes have been subjected to a good deal of statistical analysis, and if you have accurate data about past performance you can do a good deal of useful prediction. Rigging up staging or cranes to replace one bulb in the ceiling of a big industrial building is a wasteful individual operation, so it makes sense to wait until you have a number to do. While you are up there, it may make more sense to replace all the tubes, good and bad, since many of the good ones are on borrowed time. If you have good information about the average life of tubes, and the "variance" of individual life about this average, you can determine the optimum time to make your replacement. Moreover, you can predict the number that will fail before this replacement time, or the reliability of a single unit.

Suppose records are available from the manufacturer telling you that, of 100 tubes placed in service at the same time, the average number failing each week is as shown in column (2) of Table 6-2. The number surviving to the end of each week, column (3), can be derived by continuously subtracting the number failing prior to that week. You can convert this *frequency* of failures into *probability* of failure, by dividing each value by

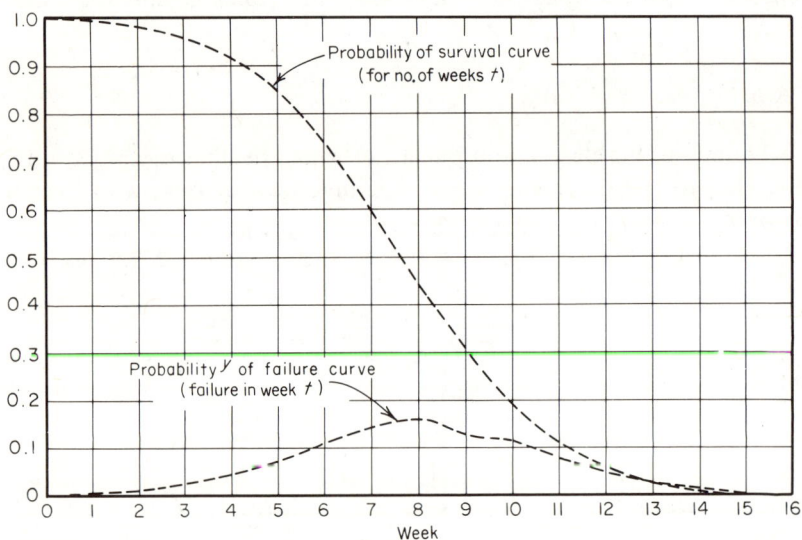

Fig. 6-9

CAPITAL EXPENDITURES

Table 6-2

(1) Week	(2) No. failing during the week	(3) Survivors at end of week	(4) Probability if it lasts to week t, that it will fail during week
0		100	0
1	1	99	1 out of 100 = 0.01
2	1	98	1 out of 99 = 0.01
3	2	96	2 out of 98 = 0.02
4	4	92	4 out of 96 = 0.04
5	7	85	7 out of 92 = 0.08
6	11	74	11 out of 85 = 0.13
7	14	60	14 out of 74 = 0.19
8	16	44	16 out of 60 = 0.27
9	13	31	13 out of 44 = 0.30
10	12	19	12 out of 31 = 0.39
11	8	11	8 out of 19 = 0.42
12	5	6	5 out of 11 = 0.44
13	3	3	3 out of 6 = 0.50
14	2	1	2 out of 3 = 0.67
15	1	0	1 out of 1 = 1.0
16	0	0	

the total number of tubes, and you can plot the probability of survival on a curve as shown on Fig. 6-9. This curve will tell you the chance any individual tube has of surviving for any period of weeks.

The fourth column of Table 6-2 tells you something else. It shows the probability that a tube *which has survived to a given week* will burn out during that week. Of the 31 which have survived to the start of the tenth week, 30 percent will burn out during the week—so that any random member of this set of 31 has a 30 percent chance of expiring during the week. In the seventh week the burnout odds are only 19 percent. You can use this column to suggest to you when the odds have got so high that you no longer wish to put up with them; instead, you replace the whole lot. (You can assume that used tubes you may find offered for sale come from populations with burnout probabilities up in this region.)

7

Market Studies and Locational Analysis

INTRODUCTION

A business, to be successful, must be well managed and must keep eternally alert to the necessity for constant improvement in its practices. But good management must be applied to a business environment that is adequate to begin with, and one of the most important elements required for success is good location. Exceptional marketing skills may bring success to a variety store located in the middle of Automobile Row, but the same skills would produce far better returns if the store had a good location working for it.

There are certain underlying factors contributing to good and bad location, and managers violate the rules at their peril. This chapter reviews some of the factors that are amenable to quantitative treatment, and that permit the manager to attach numbers to every consideration of business locations. Such treatment will enable him to say "This location is better than that location by so many dollars a year," and thus to choose on an explicit basis. It will make clear the relative importance of the separate locational elements, so that he will have a basis for choice between alternate locations.

REGIONAL ECONOMIC BALANCE

Any region in the world—a nation, a state, a city or an individual household—must balance outgo with income over the long run; if spending continues to exceed earnings it will go bankrupt.* This principle, so well understood in the abstract, often is forgotten when business or civic groups are faced with its practical application to their region.

Consider a metropolitan area of moderate size—say a city named Jonesville, with a population of 100,000, and its peripheral suburbs. It has a mix of industrial and commercial establishments, and is a regional distribution center for the small towns around it. How does Jonesville function economically?

Draw a circle around the outer fringes of the Jonesville metropolitan area. Money flows into this circle in return for goods or services sent out: goods manufactured inside the circle and exported; shoppers driving inside to patronize Jonesville stores or service establishments; individuals sending in gifts or payments; residents of Jonesville driving outside to work and returning with their salaries; and similar transactions where items of value are exported and the money payment is imported. Money flows out of this circle to pay for goods or services brought in: residents driving outside to make purchases; goods or services imported from outside; remittances sent outside for any reason; and all other transactions of this nature. In the long run, Jonesville cannot buy except with money obtained from something it sells.

Jonesville has a central business district which provides for its retail needs and serves as the principal shopping center for the surrounding region. A developer comes in and proposes that a tract bordering a major suburban approach to the city be rezoned from residential to commercial so that a large shopping center can be erected there. He cites the number of employees who will be hired in the new center, and argues that the zoning change should be approved because it will bring additional prosperity to Jonesville. How about it?

In a word—nonsense. Except as the new center attracts outside shoppers who do not come to Jonesville now, the new center will only redistribute the retail business now generated. Retail establishments are spenders, not earners; they buy goods outside to sell inside, and money must flow out to pay for these goods. Before Jonesville will support any more retail establishments, it must create a matching volume of earning industries. If the new center creates more employment for retail clerks, there will be less employment in some other category.

* A sovereign nation is no exception to this rule, but since it can create money by fiat it goes bankrupt in a different way than do states, cities, or individuals.

It may be desirable to build the new shopping center, for a number of reasons: to supplant inefficient or monopolistic merchants, to correct an unbalanced retail store mix, to bring shopping lines and methods up to date, or simply to keep pace with growing population (and hence growing earning power). But the city authorities should not authorize it under the impression that they are bringing in "new industry" or helping economic growth of the region.

PRINCIPLES OF RETAIL LOCATION

Why does the developer want to build on this major suburban approach, if overall business will not increase? Obviously because he expects business to come to *him*, though at the expense of the downtown merchants. What sort of business will he get, and how does he estimate this?

Elements of retail volume

A retail establishment in a shopping center—say a major department store in a suburban mall—can expect to generate business in the following ways:
1. *Direct draw* of consumers in its own "franchise area"—the residential areas so convenient to it that residents could shop elsewhere only at substantially more effort.
2. *Partial draw* of consumers located in areas between its site and the sites of similar retail establishments—the division being determined by relative distances, relative sizes, and relative appeal of the various centers.
3. *Interchange* with consumers who visit the area to trade at a neighboring establishment—a certain percentage of whom become impulse shoppers at the department store.
4. *Interception* of customers who pass the shopping center site en route to a larger center (such as the central business district of Jonesville)—a certain percentage of whom will be diverted from their intended destination and shop at the department store.
5. *Proportional sharing* of a somewhat larger trade area population— the increase arising from the fact that the new shopping center with its department store would make Jonesville a somewhat more attractive shopping complex to residents of the surrounding regions, and would entice some of them away from competing shopping complexes in nearby cities.

Methods 1 to 4 do not bring any new business to the Jonesville metropolitan region, but simply divert some business from other retail establishments to the new department store. Method 5, however, draws the overall boundaries of the trading area somewhat wider, so that Jonesville now has access to the additional earnings of this increment.

The above five factors will bring business to the new department store as soon as it achieves steady-state operations. There are two other factors that apply over time:

6. *Increased purchasing power* of the trading area, which comes from increase in population or from increase in income of the present population. Either of these arises from a growth in the "export" type of industry as described earlier—that type which generates goods or services inside the Jonesville trading area and sells it outside. If the population of an area is increasing, it can be assumed that this increase arises from such a growth in export industry.
7. *Development of retail competition.* As Jonesville grows geographically, and the major transportation arteries fan out from the central business district, opportunity will arise for other centers to form astride key intersections and to intercept some of the department store's business in the same way the department store did this when it was built. No locational analysis is complete unless it evaluates the probability and severity of such potential developments.

Analysis of retail volume

It is not the purpose of this chapter to describe the detailed calculations and survey techniques used to estimate the future volume of a proposed retail establishment.* However, it is useful to understand the principles on which such detailed analyses are based, and these are set forth below:

Franchise Area Market

The area within walking distance, usually taken as about $\frac{1}{3}$ mile (somewhat longer for populations where frequency of women drivers is lower than the average), approximates a captive market for an attractive and well-run shopping center. If the average household income for residents in this area and the fraction typically spent for goods of the type sold in the department store are known (both of which are available in Census statistics), it is realistic to estimate that a very high percentage of this fraction—perhaps as much as 90 percent—will come to the department

---

* A classic reference work for this purpose is Richard L. Nelson, *The Selection of Retail Locations*, McGraw-Hill Book Company, New York, 1958.

store. This assumes, of course, that no competing stores for these categories of retail goods are in or near the franchise area.

Shared Area Market

Many market studies have shown that, everything being equal, the trade area boundary between two shopping centers—the point where a resident would be indifferent about which of the two he patronized—is so located that the distances to the two centers are roughly proportional to the square roots of the attraction masses of the two centers. The "attraction mass" of a center can be expressed in terms of the total sales-floor area of all the stores that exert an individual attraction on shoppers from a distance. (A department store, furniture store, or clothing store is such a store; a drug or variety store is not, because shoppers normally do not embark on a shopping excursion of several miles primarily to visit a store in the latter category, but patronize it only incidentally to a trip stimulated by a store in the former category.) If everything is not equal—convenience, variety, attractiveness, price—these differences must be considered in arriving at an equivalent value for attraction mass.

Figure 7-1 illustrates this concept, which is based on the physical science relationship that the gravitational attraction of a body is proportional to its mass and inversely proportional to the square of its distance. If two similar centers, A and B, have attraction masses of 100 and 25 respectively (say 100,000 and 25,000 square feet), and are 6 miles apart on a residential highway, then the point of equal attractiveness, P, will fall at a distance d_a from A and d_b from B such that

$$\frac{d_a}{d_b} = \sqrt{\frac{m_b}{m_a}} = \sqrt{\frac{25}{100}} \text{ or } \frac{1}{2}$$

In this example, shoppers residing within 4 miles of A would tend to patronize A, and shoppers residing within 2 miles of B would tend to patronize B.

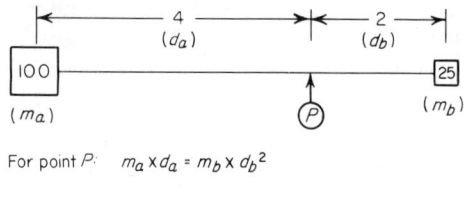

For point P: $m_a \times d_a = m_b \times d_b^2$

or $\frac{d_a}{d_b} = \sqrt{\frac{m_b}{m_a}} = \frac{1}{2}$

Fig. 7-1

MARKET STUDIES AND LOCATIONAL ANALYSIS

If there are three or more retail centers, similar calculations may be made to draw crude limits of the market area for each.

Although the above calculation utilized distance, travel time is a more relevant variable, and is the one which usually is used.

Interchange Market

Stores collect in centers for mutual reinforcement. A furniture store and a clothing store located together will do more total business than if each were alone. This is occasioned by the shopping habits of the typical consumer. A certain percentage of buyers who leave home planning to visit the furniture store (and who do visit the furniture store) will note the adjacent clothing store and will make an unplanned visit to it. The amount of such unplanned interchange shopping can run as high as 20 percent for stores that are highly compatible, and it will be negligible for stores that are completely incompatible. Indeed, total incompatibility will have a marked negative effect. A funeral home is a very poor neighbor for a variety store, because it is almost unthinkable for a customer of the former to be in a frame of mind that will induce him to drop in at the latter for an impulse purchase, but members of funeral parties will cause crowding or occupy parking which discourages purposeful shoppers at the variety store. An industrial neighbor such as an unsightly factory can blight a retail area and almost totally discourage retail shoppers. The careful selection of compatible neighbors, so that two compatible stores will enjoy a greater total volume than if they were separated, is a key element in locational analysis, and constitutes the major strategy in design of a shopping center after the overall location is determined.

What is the magnitude of this interchange effect? If two stores of good compatibility (say 10 percent) are adjacent, and the independent volume attainable by each can be estimated to be $5 million and $2 million, it can be calculated as a first estimate that 10 percent of the former's customers will call at the latter store and vice versa. This implies that the interchange dividend of the smaller store would be $500,000 and that of the larger store would be $200,000. The smaller store receives a 25 percent increase and the larger only a 4 percent increase—or they benefit inversely in proportion to the square of their relative independent volumes. While the above estimates of absolute increase may be too large (they double-count some shopping at one or the other store, because shopping done on impulse today may do away with the need to accomplish it purposefully tomorrow), this analysis highlights the tremendous benefit to smaller establishments from locations adjacent to major establishments. The tiny flower shop or newsstand located near large stores gets essentially all its business from such interchange, and would perish if thrown on its own.

Intercept Market

As a general rule, a shopper bound for a retail establishment to make a specific purchase will not pass a similar retail establishment that is on his route—he will make his purchase there. As metropolitan regions grow geographically, shopping centers develop in concentric rings at the major intersections of traffic flow to the inner centers—and intercept the large majority of all customers bound for similar facilities. The growth of central business districts in major cities is virtually frozen by this development, so that there has been almost no construction of new downtown department stores in recent years even in cities with vigorous growth. Downtown stores survive because of two principal factors: they offer wider merchandise choice, and the growth of office building population has expanded their franchise area markets.

Intercept volume calculations must estimate the traffic volume presently passing the location to shop elsewhere, must analyze the relative attractiveness of destination centers, and must estimate what percentage of this trade represents substantially the same retail choices that the center under study will offer. That done, and subject to convenient and well-marked access to the center, it can be estimated that a very high fraction of such business ultimately will be intercepted by the new center.

Such calculations measure the amount of business your center will take away from inner centers; by the same token, you must determine where the next outer ring of new centers will go when metropolitan expansion is justified, and must conclude that they will have the same effect on you.

Enlarged Trading Area Market

Just as your new establishment took part of the shared-area market from other establishments in your market area, as shown in Fig. 7-1, so will the total Jonesville retail complex, enlarged by the addition of your projected center, exercise a similar attractive force in competition with other regional centers. The result—now considering that the total Jonesville retail complex is the mass m_a shown in Fig. 7-1, and that your center has enlarged it from 100 to 144—will be to extend the limits of Jonesville's trading area, represented by point P, nearly $\frac{1}{4}$ mile farther out.

This enlargement, taking place along all access routes, will increase the total retail market available to Jonesville, and your establishment will share proportionately in this increase. It must be remembered, of course, that any similar expansion in competing regional shopping complexes will push their boundaries back closer to Jonesville.

ECONOMICS OF INDUSTRIAL LOCATION

The industrial plant appears to locate on the basis of quite different principles. Actually there is no difference in overall concept—the need to maximize overall net profit—but the factors that contribute to net profit apply to a different degree. The retail business operates under such standardized labor and material costs that net profit is almost directly dependent on gross volume, and hence locational analysis reduces to calculation of expected gross volume. Organizations doing no retail business do not require the physical presence of consumers and thus their gross volume is less dependent on plant location. The cost side of the ledger, however, is heavily dependent on location, and hence such industries must seek locations that will minimize cost.

Wholesaling and distributing

The essence of wholesaling and distributing operations is the warehousing of commodities which flow in and flow out at different frequencies and lot sizes, with profit coming from the difference between the inflow and outflow prices but reduced by the warehousing, handling, and transportation costs. Such a business may buy from manufacturers and sell to retail dealers, or it may buy from retail producers and sell to processors. An auto parts distributor is an example of the first; a country grain elevator is an example of the second.

A grain elevator is an interesting example of wholesaling, because in addition to the usual prediction problems it incorporates the problem of shipment time to maximize return in the face of seasonal prices. Grain elevators vary in storage capacity from 15,000 to 2 million bushels, and their size is related closely to the market served. A grain elevator must receive all grain presented by farmers in its market area; dry that fraction which is too moist; store the grain awaiting shipment; treat grain while in storage to prevent mildew, fungus, etc.; outload onto railcars; and select the time and destination of outgoing shipments to maximize net return.

The location of a grain elevator is a compromise of many factors. Farmers normally patronize the elevator nearest to them, so the inflow volume of a location under consideration is the volume of production of those farmers for which this location is the closest. Since this volume and the length of the harvest season for the grain under consideration* are major determinants of elevator size, a particular location cannot be chosen if the expected volume will swamp the planned capacity of the

---

* For hard winter wheat, this may be as short as 10 days, with a fifth of the total coming in a single day.

elevator to be built. Shipments must be made to the best markets, considering both transportation costs and selling price, and these must be analyzed for each location under consideration before a final choice can be made.

Manufacturing

A plant manufacturing products for sale to other industries, or for wholesale distribution, is an additional step removed from the final retail market, but still cannot ignore the location of its ultimate customers. For such a plant, the cost elements affected by location are: distance from raw materials and suppliers, production cost differentials at different plant locations,* distance from warehouses (or distributors) where final product must be sent, warehousing cost differentials at different warehouse locations, and distance from warehouses to final markets. In some businesses, where time or responsiveness is a significant item, distance from plant to final market is an important factor, but usually the industrial buyer does not place much importance on the distance to manufacturing plants of suppliers, except for shipping costs.

Locational analysis for an industrial corporation thus reduces to an exceedingly elaborate balancing process of finding the optimal solution from what may be thousands of alternate choices of plant number and location, warehouse number and location, warehouses served by various plants, and markets served by various warehouses. At the minimum (where other factors narrow the choice substantially), several different solutions should be priced out and compared. Ideally, all possible solutions should be considered and the best few presented for final choice. This is too elaborate a process to be done by intuitive or repetitive methods; it calls for the techniques of programming which are described in Chap. 10. Such techniques consist essentially of selecting a solution without regard for its optimality, and using a step-by-step approach that finds successively better and better solutions until one is found which cannot be improved. In most cases this must be done by computer, which has the speed and capacity to thread its way through thousands or millions of possible solutions and proceed systematically to the optimal one.

THE CONCEPT OF MARKET SURVEYS

No attempt will be made in this section to deal with the growing collection of techniques for doing specific market studies for particular products.

---

* This cost incorporates such factors as construction costs, labor availability, taxes, utilities, and general environmental factors.

There are, however, certain basic concepts that apply when the problem is to determine the general market for a broad category of retail products in a specific region. It is useful to understand the approach to this type of problem.

A retail merchant seeking to determine what volume he will attain if he opens an establishment in a particular location proceeds generally as follows:

1. There is an overall trade area surrounding the proposed location, within which consumers may buy from him, but outside which there is little chance they will do so. This may be a circle with his location at the center, and with a radius as small as 3 miles (for a supermarket) or as large as 25 miles or more (for a major department store).
2. This trade area can be subdivided, by spokes and concentric rings, into residential areas small and homogeneous enough so that surveys taken in each area will be reasonably applicable to the whole area.
3. The population and average income can be determined for each area, together with the fraction of such average income normally spent on the category of retail products sold by the retail merchant.
4. For each such area, a calculation can be made of the fraction of such expenditures that can be expected to be made in the proposed new establishment. This step is the heart of the survey and is a tedious but direct process. Essentially it involves listing all competing retail establishments where this category of products might be bought, assessing the relative attractiveness of each establishment to the consumers in that area (through the methods described earlier, supplemented by subjective weighting of non-quantitative factors), and splitting the potential market between all existing establishments and the proposed new establishment. This process, to be sufficiently accurate, must be supplemented by a random sampling of households in the area to determine the actual shopping habits of residents: the distribution of shopping centers they patronize at present, how they get there, and how well their present needs are met.

As a check on the above, there should be a random sampling of important competing establishments to determine the distribution of patronage to each establishment by residence areas: where the customers of this center live, how they get to the center, what establishments constitute the principal draw, and which other centers they patronize.

From this information, and using the principles described in Chap. 2, the merchant will be able to predict his future market to some specified degree of accuracy and with some known confidence in his predictions.

In an actual situation, there will be complicating factors that must be dealt with on an ad hoc basis—the effect of traffic lights to assist entry and exit at the site, visibility of the center and the individual establishment, layout, existence of traffic obstacles such as rivers or parks, and so on. On the other hand, there will be existing surveys that can ease the task considerably, such as urban renewal studies, or the studies of successful merchants.* A successful location requires the existence of sufficient potential customers, and there is no substitute for making an adequate survey to find them and identify their shopping habits.

* These generally will not be available, but their results often can be inferred from the actions of such operators. If a major chain opens a store of a certain size in a new location, it can be assumed that a market study predicted the existence of a potential volume equal to the usual standard of that chain.

8

Network Management

INTRODUCTION

The role of network management

If you were to state the essence of your managerial responsibilities, a single sentence might suffice: "I get people to achieve the aims of the organization." But listing the problems involved would take hours. There are human problems: motivation, job cognizance, discipline. There are technical problems: design, tooling, process. Certainly high on your list are organizing problems: planning to minimize lost motion, scheduling tasks to mesh smoothly, adjusting to unavoidable changes. This chapter will discuss problems of organizing.

Planning and scheduling are easy when tasks are simple and you know them well. Quantitative methods are useless when you know the best alternative intuitively. But intuition doesn't come from second sight; it represents the distillation of relevant experience, and if you lack experience in a complex situation your intuition can let you down. Experience is the key ingredient in the technique described in this chapter, but with two reservations: you are able to use experience of others as well as your own, and you don't have to find experience that fits your total project but can make do with bits and pieces gleaned from other projects.

Network management (or PERT, or critical path scheduling) squeezes from such building blocks of information the maximum guidance for controlling a complex task you haven't met before. Fed information about individual trees, it shows the shape of the forest. Its computational tricks bring order to a chaotic jumble of facts and can explore many alternatives to select the fastest or cheapest, or to assess the probabilities associated with various completion times. If the individual pieces of information you give it are precise, you get firm answers. Even when they aren't, network management often can tell you just how dependable the overall answers are. It is so much better than unaided managerial skill in planning large tasks or rescheduling around delays that major defense contractors find it indispensable. You will, too.

The concept of management control

Jot down the steps in organizing a task and you will see how network management works:
1. *You establish objectives*—because without a yardstick you can't measure progress.
2. *You plan*—by assessing the time and resources available, and estimating how much of each is required for each job of your overall task.
3. *You schedule*—by putting calendar dates on your plan, so that jobs will be done in the right sequence and no two jobs call for the same resource at the same time.
4. *You implement*—by putting the schedule to work.
5. *You evaluate progress*—with a systematic checking procedure that spots delays in time to adjust the schedule and still finish on time.
6. *You "recycle" the process*—as delays or plan changes arise, you must decide what remedial action is called for, and must predict exactly what it will accomplish, so that you adjust to circumstances *in the best possible way*.

The last five words highlight the major contribution of network management. A project of any size always contains individual jobs which can be put off for a time without causing overall delay; and from these jobs you can borrow resources to speed up critical jobs that are dropping behind. But in any complex task, it is extremely unlikely that you can find these critical jobs intuitively. Network management shows you immediately which jobs they are, and how much they can help your schedule difficulty. After you "borrow" resources, and reschedule in a way that will minimize overall delay, network management tells you when you must start each job and how many resources you must devote to it. Since you couldn't possibly see your way unaided through such a welter of overlapping jobs, network management is a must for the manager who wants to optimize his output.

THE FUNDAMENTALS OF NETWORK MANAGEMENT

What it does

In its various forms and names (PERT, PERT-Cost, PERT-Decision, Critical Path Method, etc.), the method assists you to:
1. Plan a complex job
2. Schedule the job in a workable sequence
3. Redistribute manpower to provide as nearly as possible a level work force
4. Reschedule to compensate for delays and bottlenecks
5. Determine the cost-time trade-off (what it will cost to save time)
6. Consider the economic effect of resource costs (the fixed elements of a job, equipment, and so on)
7. Determine the probability of meeting a specified date (or finding the date which you have some specified probability of meeting)

Network management is not just for huge projects which span years and occupy thousands of men. The technique came into being on such a project, but it has helped thousands of managers to improve their performance on projects of all sizes, from building an integrated steel mill to planning an office picnic. In the larger tasks, the technique enlists computer assistance; in the small ones, it calls only for pencil and ruler. Network analysis has been used with profit to:

Build a house
Make an overseas trip
Relocate a plant
Minimize the size of the permanent work force in a job shop
Gather labor statistics
Prepare a proposal for a grant
Build an overseas power plant
Prepare an advertising campaign
Build an aircraft carrier
Plan a marketing study
Shift inventory control of a business to a computer
Improve scheduling shutdowns for maintenance
Plan annual model changeover
Bring out a new product

and for many thousands of other applications.

What you must put into it

You are undertaking a project that consists of a relatively large number of individual jobs. Network management imposes its own discipline on you, in that there are certain things you must know about your task. These are things a good manager ought to know anyway, before he starts a project; frequently he does not, but counts on muddling through in hopes that subordinates will provide the missing ingredients. Network management forces him to face up to his planning needs before he starts, producing some management improvement from this factor alone.

You are asked for three things:
1. *Time estimate for each job:* This should consist of a "most likely" time estimate for doing the job in the normal way. (Additionally, if you want the probability of meeting your schedule, you should include an "optimistic" time estimate that assumes you get the breaks, and a "pessimistic" time estimate that assumes everything goes wrong.)
2. *Sequence of jobs:* This should record what jobs must be completed immediately before each job you are scheduling, and what jobs will follow immediately after.
3. *Cost of speeding up each job:* This consists of the cost of shift work or overtime, or of putting on an inefficient shift size for a job—the cost of what it takes to do the job faster than normal.

Diagramming rules

The standard PERT diagram is a graphic picture of your project. If you were to PERT the very rudimentary project of getting yourself to work in the morning, it would diagram like Fig. 8-1.

This network diagram consists of "activities" and "events." An *activity* is an individual job or task, shown here by an arrow labeled with the duration of the job. An *event* is the start or finish of an activity, shown here by a circle. The activity, "shave" can be identified either by the letter A or by ①-②. To avoid unnecessary duplication, the letters can be omitted and each activity identified by the event numbers at tail and head of the arrow.

How long will it take you to get to work? There are three paths to check:

①-②, ②-④, ④-⑥, ⑥-⑦ takes 10 + 15 + 20 + 15 or 60 minutes
①-③, ③-④, ④-⑥, ⑥-⑦ takes 5 + 15 + 20 + 15 or 55 minutes
①-⑤, ⑤-⑥, ⑥-⑦ takes 30 + 20 + 15 or 65 minutes

NETWORK MANAGEMENT

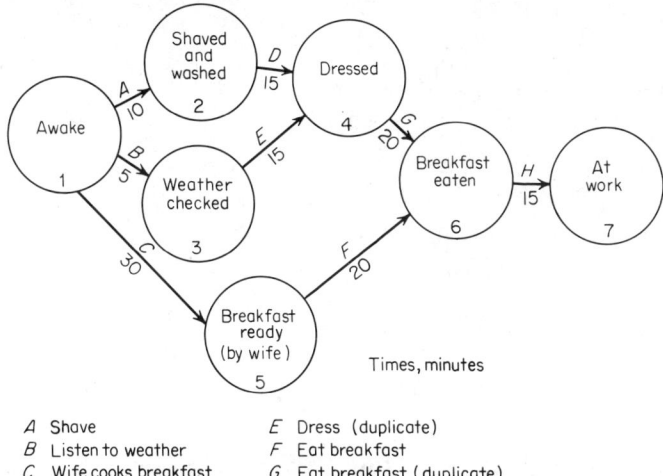

A Shave
B Listen to weather
C Wife cooks breakfast
D Dress
E Dress (duplicate)
F Eat breakfast
G Eat breakfast (duplicate)
H Ride to work

Fig. 8-1

The largest of these, 65 minutes, is the "critical path" for the operation. This process tells you the following useful things:

1. Any job on the critical path is "critical" to your schedule in that it delays the entire project.
2. The overall project will not be shortened if you shorten the duration of a job not on the critical path.
3. "Slack time" exists for jobs not on the critical path, making them useful sources for manpower or other resources to put on some "critical" job. (This talk of transferring manpower and resources applies to something a bit more complex than your getting-to-work schedule, but the idea is the same.)

THE COST-TIME TRADEOFF

A simple one-job case

Doing a job faster than its "normal" time usually costs money, and the more time saved the greater the cost. Imagine that you are having a room plastered, and the job amounts to 12 man-days. Since plastering is done most effectively with a two-man team (you decide), and since day-shift rates are lowest (say $6 an hour, compared with $6.60 for second shift and $7.20 for third shift), you plan for normal—or least-cost—accomplishment by using 2 men per day for 6 days, day shift only, or:

Six-day (Normal Time) Job

    12 man-days at 2 men per day: 6 days
    12 man-days at $48 per day: $576

If you want to shorten the time to 4 days, your only alternative is to work the job on the day shift for the 4 days, and make up the remainder of the second shift at $52.80 per day ($6.60 per hour).

Four-day Job

    8 man-days on day shift at 2 men per day + 4 man-days on second shift at 2 men per day   4 days
    8 man-days on day shift at $48 per day + 4 man-days on second shift at $52.80 per day $595.20

If you want to shorten the time to 2 days, you see that all three shifts are needed:

Two-day Job

    4 man-days on day shift at 2 men per day + 4 man-days on second shift at 2 men per day + 4 man-days on third shift at 2 men per day   2 days
    4 man-days on day shift at $48 per day + 4 man-days on second shift at $52.80 per day + 4 man-days on third shift at $57.60 per day $633.60

Since further speedup would involve wastefully crowding too many men on the job, you take 2 days as the "crash" minimum for this job with any reasonable economy.

A simple graph showing the cost-time trade-off would look like Fig. 8-2. The solid line represents the range of cases you considered. Below 2 days the cost rises rapidly because of wasteful crowding, and above 6 days the cost rises somewhat because a one-man plastering operation is less efficient. The cost to save 4 days is $57.60—the difference between $633.60 and $576—or approximately $14.40 per day saved. (Not exactly, of course, because the line on the graph is not exactly straight between the 2-day and 6-day points, but it is close enough.)

A simple multijob case

The above calculation tells you how you would determine two things:
    1. The cost per day to shorten the job, from "normal" to "crash" time ($14.40 per day)
    2. The normal and crash durations of the job (6 and 2 days, respectively).

NETWORK MANAGEMENT

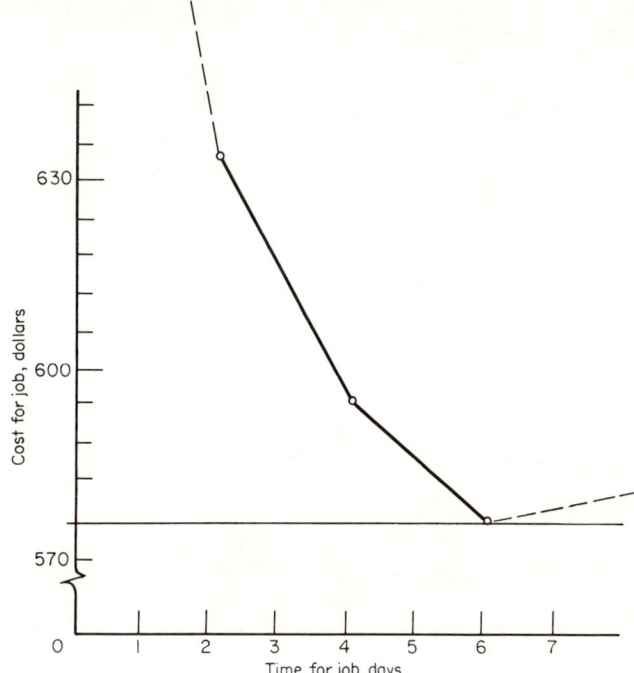

Fig. 8-2

Now try a project consisting of several such jobs, for each of which you have determined previously the *normal time*, the *crash time*, and the *cost per unit time saved*.

Table 8-1 gives these estimating factors for each job, and Fig. 8-3 is the network showing the job relationship, with the factors from Table 8-1 written in. You see by inspection that the normal time is 11 days, along critical path ①-②, ②-③, ③-④, and the crash (shortest feasible) time is 8 days along path ①-③, ③-④.

This project can be done in 11 days for $310; anything shorter will

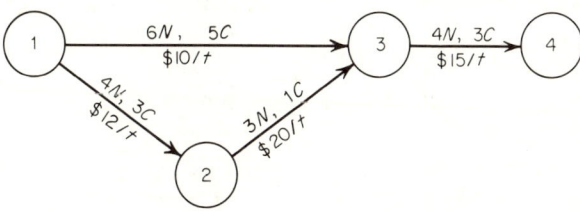

Fig. 8-3

Table 8-1

Job	Normal time, days	Crash time, days	Normal (minimum) cost	Cost per unit-time reduction
①–②	4	3	$100	$12
①–③	6	5	60	10
②–③	3	1	80	20
③–④	4	3	70	15
			$310	

cost you money. You would like to find project costs for the various completion times between 11 and 8 days:

Normal Schedule (11 *Days*)

    Critical path: ①–②   ②–③   ③–④
    Normal time: 4 + 3 + 4 = 11 days
    Normal cost: $100 + $80 + $70 = $310
                             + $60*

10-day Schedule

Where can you save a day most cheaply? Clearly at event ①–②, where it costs only $12 per day. You tabulate the results of shortening event ①–② from 4 to 3 days:

    Critical path: ①–②   ②–③   ③–④   (unchanged)
    New time: 3 + 3 + 4 = 11 days
    Cost increase: $12
    10-day cost: $310 + $12 = $322

9-day Schedule

You have shortened the least-cost event all you can, so you must look at other events. Event ③–④ is the cheapest event with time available, so you tabulate the results of shortening it from 4 to 3 days:

    Critical path: ①–②   ②–③   ③–④
    New time: 3 + 3 + 3 = 9 days
    Cost increase: $15
    9-day cost: $322 + $15 = $337

* Cost for job ①–③.

8-day Schedule

Proceeding similarly, you shorten event ②-③ and tabulate the results:

Critical path:	①-②	②-③	③-④	
New time:	3 +	2 +	3	= 8 days
Cost increase:	$20			

But when you shortened this path to 8 days, it ceased being the critical path; since ①-③, ③-④ requires 9 days, event ①-③ must be shortened at the same time, with the following results:

Critical path:	①-③	③-④	
New time:	5 +	3	= 8 days
Cost increase:	$10		
Total 8-day cost:	$337 + $20 + $10 =		$367

No further reduction is feasible, and you therefore tabulate the results on Fig. 8-4 and take a look at them. The graph shows clearly

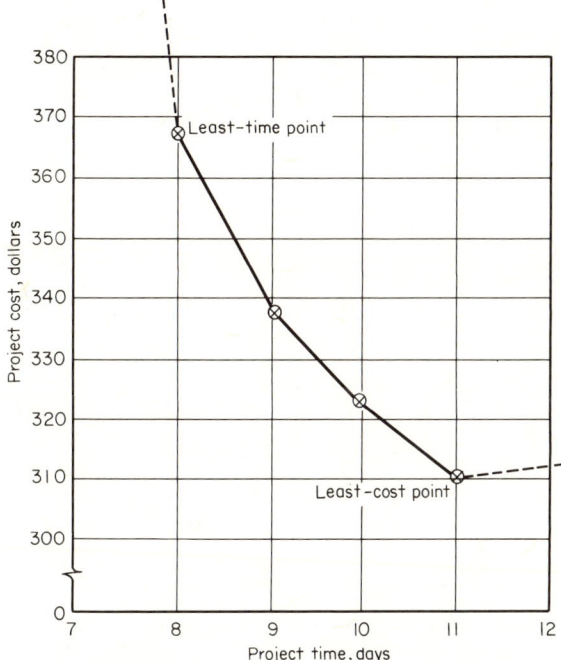

Fig. 8-4

Table 8-2 Project cost-time curve, linear program computer-run installation of gas cracking furnace. Project duration time: 248 days*

Job code	Sequence code	Description	Job duration	Cost	Earliest Start	Earliest Finish	Latest Start	Latest Finish	Total Float	Free Float
1001	1 2	Lead time	44M	$	0			44	0	0
1114	1 16	Field start restraint	84M		0	84	89	173	89	68
1014	2 3	Tube bend drawings	22M	710	44			66	0	0
1007	2 4	Furnace design drawing	22M	710	44			66	0	0
1113	2 16	Dummy	0M		44	44	173	173	129	108
1005	2 19	Building power and lighting drawings	21M	680	44	65	192	213	148	0
1104	3 4	Dummy	0M		66			66	0	0
2010	3 27	Procure tubes and bends	30M	18,000	66	95	193	223	127	127
1009	4 5	Pipe arrangement drawings	43M	1,380	66			109	0	0
1006	4 6	Equipment foundation drawings	33M	1,050	66	93	97	130	31	10
1008	4 7	Furnace detail drawings	22M	710	66	88	87	109	21	0
1103	5 6	Dummy	0M		109	103	130	130	21	0
1101	5 8	Dummy	0M		109			109	0	0
1010	5 12	Piping details	33M	1,050	109	142	130	163	21	0
1011	5 22	Instrument drawings	43M	1,380	109	152	110	153	1	0
2008	5 25	Valves, fittings	30M	4,000	109	139	188	218	79	54
2001	5 29	Insulation procurement	5M	3,000	109	114	233	238	124	94
1002	6 13	Trench drawing	22M	710	109	131	130	152	21	0
2013	6 16	Foundation steel	30M	8,000	109	139	143	173	34	13
1100	7 8	Dummy	0M		88	88	109	109	21	21
1102	7 9	Dummy	0M		88	88	143	143	55	0
1013	7 10	Refractory drawings	11M	355	88	99	152	163	64	0
2004	7 17	Structural steel procurement	30M	17,000	88	118	153	183	65	65
2007	7 27	Stack and quencher procurement	40M	6,000	88	128	183	223	95	95
1003	8 11	Steel drawings	54M	1,720	109			163	0	0
2006	9 17	Furnace steel procurement	40M	33,000	88	128	143	183	55	55
2011	10 24	Refractory procurement	30M	18,000	99	129	163	193	64	64
2003	11 17	Pipe support procurement	20M	5,000	163			183	0	0

132

Activity	i	j	Description	Duration	Cost ($)	ES	EF	LS	LF	TF	FF
1012	12	20	Procure power and lighting drawings	22M	710	142	164	201	223	59	0
2009	12	24	Fabricated pipe procurement	30M	4,000	142	172	163	193	21	21
1004	13	14	Trench grating drawing	11M	365	131	142	152	163	21	0
2002	14	15	Trench grating	10M	2,000	142	152	163	173	21	0
1112	15	16	Dummy	0M		152	152	173	173	21	0
3001	15	25	Trench and trench grating	5M	1,950	152	157	213	218	61	36
3007	16	17	Equipment foundations	10M	7,800	152	162	173	183	21	21
3002	16	30	Storage part details	5M	2,325	152	157	238	243	86	86
3004	17	18	Structural steel shell pipe supports	10M	2,600	183			193	0	0
1105	18	21	Dummy	0M		193	193	233	233	40	0
1108	18	23	Dummy	0M		193			193	0	0
1111	18	28	Dummy	0M		193	193	233	233	40	40
2005	19	21	Power and lighting procurement for building	20M	5,000	65	85	213	233	148	108
2012	20	21	Process power and lighting procurement	10M	5,000	164	174	223	233	59	19
3011	21	30	Power and lighting	10M	2,080	193	203	233	243	40	40
2014	22	23	Instrumentation procurement	40M	6,000	152	192	153	193	1	1
1107	23	24	Dummy	0M		193			193	0	0
3010	23	30	Instrumentation	15M	8,000	193	208	228	243	35	35
1106	24	25	Dummy	0M		193	193	218	218	25	0
3005	24	27	Refractory installation	30M	16,000	193			223	0	0
3009	25	26	Piping erection	15M	10,000	193	208	218	233	25	0
1109	26	28	Dummy	0M		208	208	233	233	25	25
1110	26	29	Dummy	0M		208	208	238	238	30	0
3006	27	28	Stacks and quenchers	10M	6,400	223			233	0	0
3006	27	30	Tubes and bends	20M	6,000	223			243	0	0
3003	28	30	Painting	10M	2,940	233			243	0	0
3012	29	30	Insulation	5M	1,500	208	213	238	243	30	30
3013	30	13	Cleanup	5M	1,200	243			248	0	0
			Total cost		$214,315						

* Reproduced from *Critical Path Technique*, (1961 ed.) courtesy of Catalytic Construction Company, Philadelphia, Pa.

that the cost for a unit of time saved keeps increasing. This is a trivial example, but imagine it is too big for you to have seen your way through the network by eye, and you see the usefulness of a technique that presents you the clear-cut choices for your decision. If you know what time is worth to you, you can make the decision.

Many contracts contain cost penalties for time overruns, and some contain bonuses for time saved. If you are to make a rational decision on how to schedule your project—where to put the shift work or overtime, when to shift resources from one job to another—you need facts. Network management gives you facts.

An actual multijob case

You don't undertake the above calculations by hand, of course. The computer does the work for you, using one of many available problems that can provide answers like those above before you can say "canned program." Table 8-2 shows a normal time sequence for a small 56-job project—installation of a gas cracking furnace. This is the minimum-cost way to install the furnace, and the computer calculates a 248-day project time at a cost of $214,315. Now you would let the computer try shorter times precisely as you did above, and compute the cost of each. The results of a series of these are shown below:

Project time duration, days	Project cost	Cost per day saved
248	$214,315	
233	214,555	$ 16
225	214,794	30
193	219,517	148
181	223,900	366
163	233,433	530

To shorten the project from 248 to 233 days costs only $16 per day, or $240 total—a trifling cost to pay for saving 15 days on a project of this size. The decision-maker doubtless will elect to buy this time. To shorten it from 181 to 163 days costs $530 a day, or about $9,500; the decision on whether to buy this time will not be so simple.

Total cost analysis

How does the rational executive make this decision? The above analysis covers project cost only, and there are other costs to be considered.

NETWORK MANAGEMENT

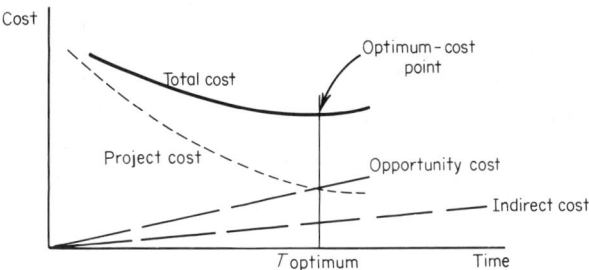

Fig. 8-5

You can group the bulk of these other costs into two broad categories:
1. *Indirect costs*—overhead costs that vary pretty much *with time*, such as equipment rental, guards, depreciation, and timekeepers
2. *Opportunity costs*—the revenue or benefit that you will start earning after the project is completed, but must forego when you extend its completion date*

You can estimate, for each possible time duration of a project, the project cost, indirect cost, and opportunity cost. The project cost decreases with increasing project time, as the previous section shows, and at an ever-diminishing rate of decrease—in other words, a curve. The indirect and opportunity costs will increase with increasing project time, on something close to a straight-line basis. For any given project time duration, the sum of the three costs will give you total cost. Figure 8-5 shows these costs as they might appear in a typical project. You can see that, although project cost is decreasing and the other two costs are increasing, there is a project time for which total cost is a minimum.

NETWORK MANAGEMENT ALGORITHMS

How to introduce a new product

This section really won't tell you how to introduce a new product in your business, because you're the one who knows the necessary steps and their interrelationships. But if the individual steps, and their interrelationships and time durations were as indicated in Table 8-3, then the corresponding network would be as shown in Fig. 8-6. (In the interest of clarity, the example uses normal times only; in a real case you might find

---

* Opportunity cost exists whether the project will take in revenue or not, because any project you undertake is being accomplished to achieve some gain or avoid some loss—and ultimately you ought to be able to place a dollar value on this, or else did not act rationally in undertaking it in the first place.

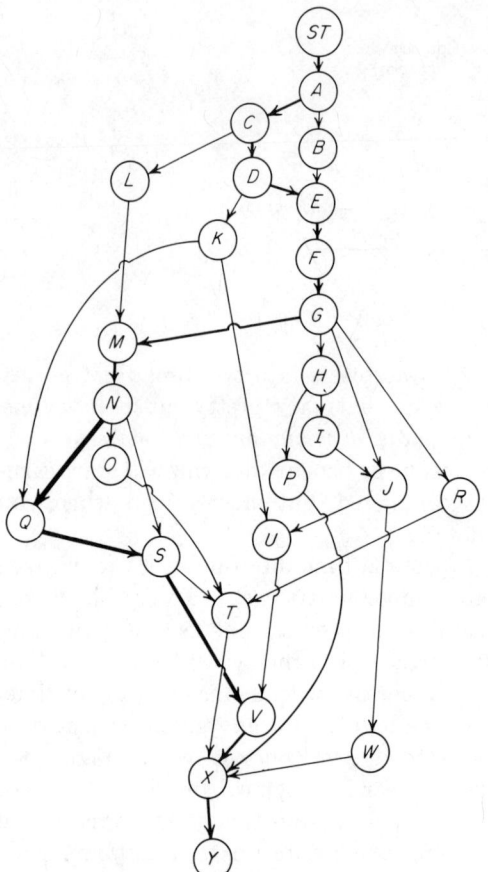

Fig. 8-6 New product introduction (network).

it profitable to include the cost-time trade-off as discussed earlier.) You can see that Table 8-3 and Fig. 8-6 contain the same job routing; the only difference is that one is tabular and the other graphic.*

If you were to plot every possible path through the network shown on Fig. 8-6, you would find that there were 29 of them. The shortest path is ABEFGHPXY and requires 14 weeks; the longest is ACDEFGMNQSVXY and requires 32 weeks. Even in this simple

---

* You may notice something else: the circle (or "node") now represents the event itself, rather than the start or finish of an event. This is a change in the ground rules we stated at the start of this chapter, but you must stay flexible because networks are drawn using a number of different conventions. The one we described at the start is the earlier technique, but that used in Fig. 8-6 has some advantages that recommend its use.

NETWORK MANAGEMENT

Table 8-3 New product introduction (job tabulation)

Step	Predecessor job(s)	Description	Duration, weeks
A		Approve new product study	1
B	A	Develop plan for market introduction	2
C	A	Prepare product drawings	3
D	C	Test, modify, and approve packages	2
E	B, D	Approve unit costs	1
F	E	Distribute sample and questionnaire in field	3
G	F	Evaluate questionnaires	2
H	G	Develop, review, and select final container, carton and shipping box	2
I	H	Test and approve container, carton, shipping box	1
J	G, I	Plan media, advertising schedule, and sales literature	4
K	D	Develop manpower needs and hire and train employees	6
L	C	Design and develop production equipment	3
M	L, G	Manufacture and install equipment	7
N	M	Debug equipment	2
O	N	Design and establish production and quality control standards	2
P	H	Make engraving plates and manufacture labels, shipping boxes, and cartons	1
Q	N, K	Order, receive, and inspect raw materials	5
R	G	Prepare and inspect warehouse space	4
S	N, Q	Conduct final trial run	1
T	S, R, O	Manufacture new product, first production run	2
U	J, K	Conduct sales meetings	1
V	U, S	Call on customers	3
W	J	News release—new product in demand	1
X	V, T, W, P	Process orders and ship product against orders	2
Y	X	Finish	0

project it takes too much calculating to find all these paths and select the critical path. In a project of any size the task would be prohibitive. You need an "algorithm," which is the name operations researchers give to a systematic step-by-step solution technique.

The "earliest-finish" algorithm

A very simple project will illustrate this method: the manufacture of two machined parts which will be called X and Y. Preparing the network requires you to:
1. List the individual jobs.
2. Put a time estimate on each.

3. Indicate, for each job, the job or jobs that must be completed immediately before it can start (the "predecessor" jobs). This gives you Table 8-4 which contains all the information you need to sketch the network of Fig. 8-7.

The algorithm for finding the earliest time that the project can finish is quite direct:*

1. Using Fig. 8-7, write "start time" for each job on its left and "finish time" on its right.
2. Start the project at zero time by putting the start time of 0 in the space to left of the first event.
3. Since this event takes 0 minutes to complete, write the finish time of 0 in the space to its right.
4. Event b can start as soon as event a is completed; write a start time of 0 to the left of b.

* Gerald L. Thompson and his colleagues first published this specific algorithm; other techniques have been used by other workers in the field.

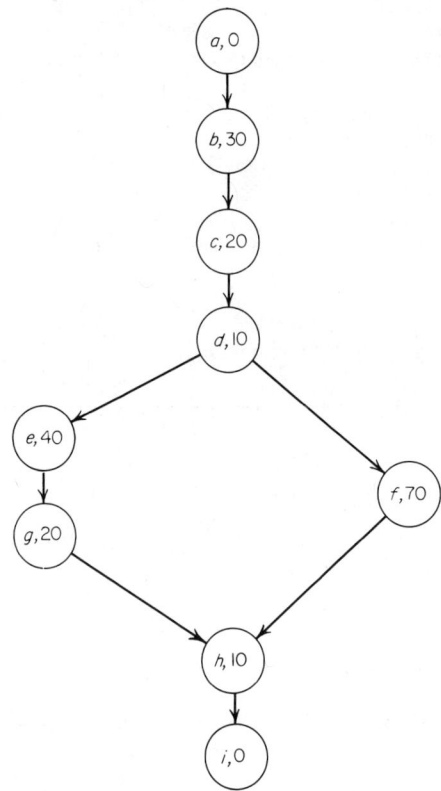

Fig. 8-7

NETWORK MANAGEMENT

Table 8-4

Job	Description	Predecessor job(s)	Time, minutes
a	Start		0
b	Make pattern	a	30
c	Cast X and Y	b	20
d	Cut X and Y apart	c	10
e	Machine X	d	40
f	Machine Y	d	70
g	Drill and tap X	e	20
h	Assemble X and Y	f, g	10
i	Finish	h	0

5. Now add the 30 minutes event b takes, to make a finish time of 30 for b.
6. Whenever there are two or more preceding events (as f and g preceding h), take the latest finish time of any preceding event as the start time for the following event.
7. Continue the above procedure to the end.

Figure 8-8 shows this completed procedure; the earliest finish is contained in the box on the right of the final job: 140 minutes. For consistency, you should label all start times on Fig. 8-8 "earliest start" and all finish times "earliest finish."

The "latest-start" algorithm

It is unnecessary to describe this algorithm in detail, since it simply operates in reverse of the previous algorithm. A finish time is selected for the project and entered to the right of the final event, and you back through the network by subtracting event times until you arrive at the latest time you can start the first event in order to meet your selected finish time. In this example, if you select any finish time earlier than 140 minutes you will require an impossible start time. If you select a latest finish (L.F.) of exactly 140—the same as your earliest finish (E.F.) —you will learn some useful facts about the network.

In Fig. 8-9 the latest start (L.S.) and L.F. for each event have been added in bold type under each of the corresponding boxes in Fig. 8-8. (It would be well for you to trace this sequence of "latest" times back through the network yourself, starting with the L.F. of 140 to the right of the final event, to make sure you are clear about the procedure.)

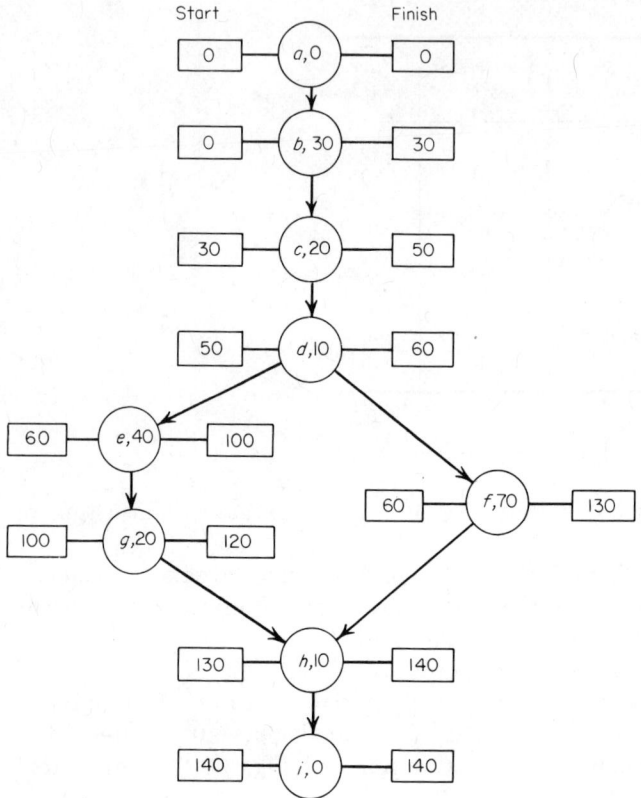

Fig. 8-8 Earliest-finish algorithm.

The critical path

It is easy to spot the events of Fig. 8-9 that lie on the critical path: simply those for which the earliest start (E.S.) and the L.S. are the same. Don't forget, however, that you arbitrarily decided your L.F. would be the same as your E.F.: 140. If you had permitted your L.F. to be 165 instead of 140—25 minutes later than the earliest possible time—the events on the critical path would have been *all those for which the L.S. was 25 minutes later than the E.S.* And all the events *not* on the critical path would have had a L.S. *more than* 25 minutes later than the E.S.

Slack

Take a look at event e. The earliest you can start this job is 60 minutes (because event d isn't finished until then). If you wish, however, you can delay the start of event e until 70 minutes (which delays the start of event g from 100 to 110), and you still will start event h at 130—the earliest the completion of event f would let it start anyway. You have 10 minutes leeway, then, in the starting time for event e; or in network parlance, it has 10 minutes, "total slack"—defined simply as the difference between E.S. and L.S. (or, of course, between E.F. and L.F.).

Why is it called "total slack" instead of just "slack"? Because there is more than one kind of slack, depending on how free you can feel in spending it. Take a look at event g, and you see that it, too, has 10 minutes total slack. You could delay e or g as much as 10 minutes without delaying the overall project. But if you use all 10 minutes at e, delaying its start to 70 and its completion to 110, you erase the slack you had at g. If you use the 10 minutes at g, however, you do not affect any following event. This difference defines a second form of slack known as "free slack," characterized by the fact that if you do not use it on the event at hand it expires and is lost forever. More formally, the free slack of an event is *the amount it can be delayed from its earliest start without*

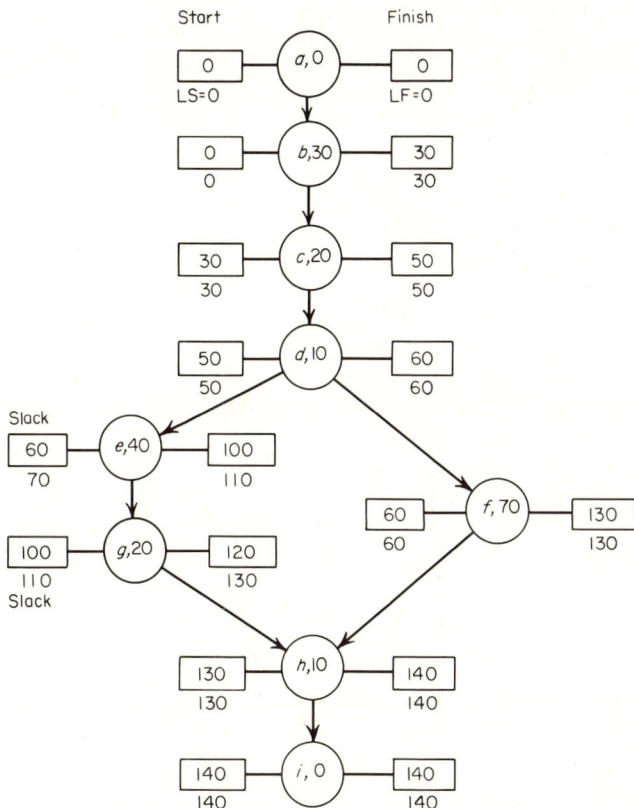

Fig. 8-9 Latest-start algorithm.

affecting the earliest start of any follow jobs. The free slack for event e is zero, but for event g it is 10 minutes.

There is a third form known as "independent slack," which looks at interference with preceding events as well as follow events. Much less important in analysis (practically, you can ignore it), the independent slack of an event is the amount the event can be moved in time without affecting either the earliest start of any follow job or the latest finish of any preceding job.

Algorithms applied to new product example

When you analyzed Fig. 8-6 by trial and error, you found that a critical path of the project took 32 days—but you had to trace out all possible

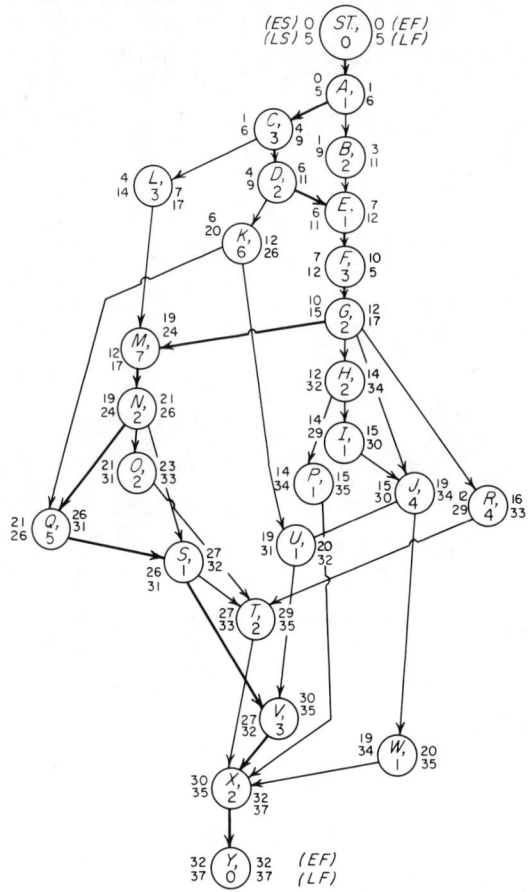

Fig. 8-10 New product introduction—earliest-finish and latest-finish solutions.

paths to be sure which was the longest. Figure 8-10 uses the algorithms described above to determine the critical path directly, and to calculate slack times as well. Note that, although the earliest finish is calculated to be 32 days, the latest finish of the project is selected arbitrarily as 5 days later (presumably because management didn't require its completion until the later date). The critical path (heavy lines on Fig. 8-10) can be found by drawing a line through each event whose latest start is exactly 5 days later than its earliest start. All the other events have some total slack, but only certain of them have free slack.

Why do you want to identify the free slack events? Because they are the events that can be delayed with impunity—without affecting any subsequent events. And they constitute the most logical place to look for resources to make up delays occurring elsewhere in the project. The manager of a complex project seldom knows with certainty all the implications of delaying one job or another; often his tendency is to borrow from jobs with completion dates furthest in the future, but with no assurance that these jobs are not themselves critical and thus simply compounding his problem. Network management permits him to borrow from the jobs he knows can stand it the best and to make his decisions in a systematic way.

MATRIX SOLUTION METHOD

Triangular matrix of activity times

The critical path drawing with its circles and arrows is a clumsy device to manipulate, and furthermore the computer doesn't like it. A neater and quicker system is to express the same relationships on a grid, or matrix format, and carry out all the computations on this matrix. The machine shop network of Fig. 8-7 can be represented in matrix form as shown on Fig. 8-11. Each figure in a cell of the matrix is identified by a letter in the left column ("From") and a letter in the top row ("To"); the number 30 thus represents the time required to go *from completion of event a to completion of event b.*

Earliest-finish algorithm

Note the vertical grid running down the left of the matrix. You will fill in these cells with the E.F. times for each event, with the same values you entered in the right-hand blocks of Fig. 8-8. Try it, not looking at the network drawing, but using only the information you can obtain from

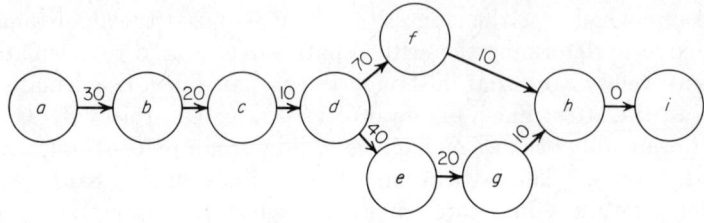

From\To			a	b	c	d	e	f	g	h
	0		a	30						
	30		b		20					
30+20	=50		c			10				
50+10	=60		d				40	70		
60+40	=100		e						20	
60+70	=130		f							10
100+20	=120		g							10
130+10 or 120+20	=140		h							

	a	b	c	d	e	f	g	h
	0	30	50	60	110	130	130	140

Fig. 8-11 Triangular matrix of network times.

looking at the matrix:

Row a entry: Enter 0
Row b entry: Only route to b is from a, so enter 30
Row c entry: Only route to c is from b, so add 20 to b entry = 50
Row d entry: Only route to d is from c, so add 10 to c entry = 60
Row e entry: Only route to e is from d, so add 40 to d entry = 100
Row f entry: Only route to f is from d, so add 70 to d entry = 130
Row g entry: Only route to g is from e, so add 20 to e entry = 120
Row h entry: Two routes to h $\begin{cases} \text{from f, so add 10 to f entry} = 140 \\ \text{from g, so add 10 to g entry} = 140 \end{cases}$
                                                  Earliest you can start h is
                                                                          latest of these = 140

     Earliest finish = 140

Latest-finish algorithm

Note the horizontal grid running across the bottom of the matrix. You will fill in L.F. times, working backward from the right cell. Select 140

NETWORK MANAGEMENT

arbitrarily as the L.F. you can accept, and proceed:

Column h entry:	Enter 140
Column g entry:	Only route back to g is from h, so subtract 10 from h entry = 130
Column f entry:	Only route back to f is from h, so subtract 10 from h entry = 130
Column e entry:	Only route back to e is from g, so subtract 20 from g entry = 110
Column d entry:	Two routes back to d $\begin{cases} \text{from f, so subtract 70 from f entry} = 60 \\ \text{from e, so subtract 40 from e entry} = 70 \end{cases}$ Latest we can finish d is *earliest* of these = 60
Column c entry:	Only route back to c is from d, so subtract 10 from d entry = 50
Column b entry:	Only route back to b is from c, so subtract 20 from c entry = 30
Column a entry:	Only route back to a is from b, so subtract 30 from b entry = 0

The results of these two algorithms are entered on Table 8-5.

The critical path

Bring down the E.F. grid and turn it sideways so you can write its values directly under the corresponding values of the L.F. grid and compare them more conveniently. (In matrix algebra procedure, you have transposed a column vector to a row vector.) Since the E.F. and L.F. in the last cell are the same—each is 140—you can identify the events on the critical path as those where the two values also are the same. (If the L.F. had been 10 minutes later than the E.F. in the last cell—and this is a value the decision-maker selects for external reasons—then each event on the critical path would have shown a similar difference of 10 minutes between E.F. and L.F.; all events not on the critical path would have shown a difference larger than 10 minutes.)

All events except e and g are, by this test, on the critical path.

The project schedule

You have calculated only the finish time for each event, but it is a simple matter to subtract the duration of each event from its finish time to get

Table 8-5 Network analysis results

Job	Time	E.S.	E.F.	L.S.	L.F.	T.S.	F.S.
a-b	30	0	30	0	30		
b-c	20	30	50	30	50		
c-d	10	50	60	50	60		
d-e	40	60	100	70	110	10	
d-f	70	60	130	60	130		
e-g	20	100	120	110	130	10	10
f-h	10	130	140	130	140		
g-h	10	130	140	130	140		

the starting time for that event. In the schedule of Table 8-5 this has been done to get E.S. from E.F. and L.S. from L.F., for each event.

You have one more set of entries to compute—those for total slack and free slack. The former is clear-cut: for each sequence it is simply the difference between E.F. and L.F. The determination of free slack for each sequence involves the same logic you used on page 140, as follows:

Look at sequence d-e. Its earliest finish is 100 and latest finish is 110, so its total slack is 10, and thus its free slack can be no greater than 10. But the earliest start of the following sequence, e-g, is 100, so you cannot delay sequence d-e at all past its earliest finish of 100 without delaying the start of the following sequence. Thus its free slack is 0.

Look at sequence e-g, the other one with some total slack. Doing it in symbols this time, you have:

$\text{E.F.}_{eg} = 120$, $\text{L.F.}_{eg} = 130$, so $\text{T.S.}_{eg} = 10$
$\text{E.S.}_{eg} = 130$
$\text{F.S.}_{eg} = \text{E.S.}_{gh} - \text{E.F.}_{eg} = 130 - 120 = 10$*

Triangular matrix for new product introduction

The triangular matrix of activity times for the new-product example of Table 8-3 and Fig. 8-6 is shown in Fig. 8-12.

ADJUSTING TO DELAYS AND CHANGES

The schedule resulting from your network analysis, such as that shown for the simple example of Table 8-5 or that for the gas cracking furnace installation of Table 8-2 holds good only until there is some unanticipated delay in one or more activities or there is some change in your required dates. At this time, if you do not modify the schedule, it is as outdated as yesterday's newspaper. But it is precisely the ability to present meaningful alternatives, so that your schedule may be revised most effectively, which constitutes the greatest advantage of network analysis.

Look at Table 8-5, and imagine that each job in this trivial example is a sizable task involving many man-days. Suppose you have dropped behind on job b-c, and must find a way to make up your schedule. You can, of course, pile on overtime—but overtime is expensive. A better solution is to see where else in the operation there is a job with slack: one that is not on the critical path and thus can afford to loan manpower and accept the resultant delay *without delaying the overall project*. In Table

* But never more than T.S._{eg}.

NETWORK MANAGEMENT 147

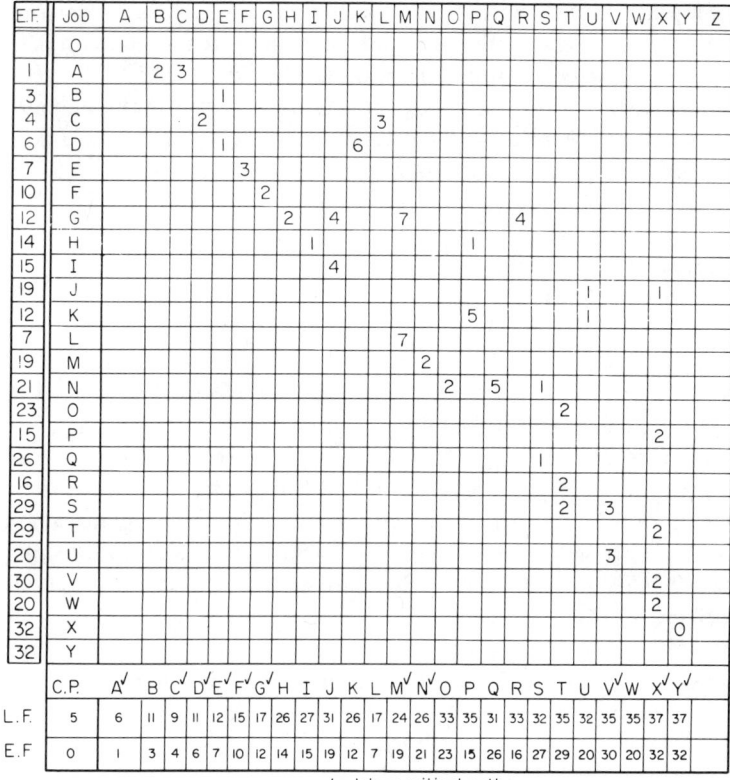

Fig. 8-12 Triangular matrix of activity times.

8-5, job e-g is such a noncritical job, for the right-hand column shows that it has some free slack.

Table 8-2 has a similar right-hand column, headed "Free Float" (the same thing as free slack). Job 2010, covering procurement of tubes and bends, has 127 days of free float; clearly this job can furnish manpower and be delayed a good long time without hazarding the overall project. It is most helpful to the project manager to have this sort of information immediately and continuously available. If this were a larger task, with several noninterchangeable trades involved, the jobs would be coded by departments, and the project manager in trouble on a given job would look for slack jobs involving the same trade.

All right. You've borrowed from the job with slack and rescheduled it accordingly. Now put your rescheduled jobs back into the computer, and read out an entirely new computer run such as that of Table 8-2. You can see at once the total implications of the change in schedule you are making, *before* you perform the work.

It has been popular, with firms making superficial use of network management, to draw the initial network diagram, but not to keep it up to date as schedules encountered inevitable changes. This "showcase PERT" is no good. There are few spectacles as pathetic as the manager who shows visitors his beautiful network schedule but can't relate its activities to anything currently happening in the plant.

OTHER POTENTIALITIES OF NETWORK MANAGEMENT

It is not always enough to have the best estimate of when a project will finish. You may need something in the way of guarantees. Your new office building is scheduled for completion in 24 months, but when is it safe to give up the lease on your present offices? If 24 months is your best estimate, this really means there is a 50 percent chance of beating this date *and a 50 percent chance of missing it*. This is no good. You need to know a date within which you can be 90 percent sure of finishing, or perhaps 99 percent—and retain the lease on the old offices until this date, if it doesn't cost too much. There comes a break point where the monthly cost of rental on your old building is more than you want to pay as insurance against, say, a 1 percent chance of missing the date on your new offices—and this is the point at which to terminate your lease.

Probability information of this sort is the raw material with which good managerial decisions are made. If you want your network schedule to give answers in probability form, you must provide individual job estimates in the form of "most likely," "optimistic," and "pessimistic" completion dates; the computer program then makes use of the confidence interval concepts described in Chap. 2 to compute a completion date corresponding to any degree of certainty you specify. (Except 100 percent certainty; the job duration for that is infinity.)

Workload smoothing

When you make a critical path schedule, such as the new product introduction of Fig. 8-10, your aim is to determine the earliest possible finish, without regard for manpower requirements. For relatively small projects this is fine, but if you are scheduling a major task you need to pay more attention to providing a level workload for your work force. A schedule calling for 1,000 electricians this week and 300 the next will bankrupt you.

Computer programs exist for handling this problem. They start with your first schedule, made without regard to manpower requirements, which results in the *earliest possible* completion date. You determine how much delay in project completion date will be acceptable to you. The computer locates the part of the schedule where manpower requirements peak, and by trial and error selects jobs to be delayed so that the peak gets leveled off but the final project completion date is met. It keeps doing this, peak by peak, until manpower needs are leveled as much as the completion date requirements will permit.

The future

Other applications have been investigated, such as PERT-Reliability, PERT-Decision, and so on, to extend this versatile technique to make still greater contributions to management. The essence of all of them, however, is providing a generalized system into which the manager inserts incremental data and decisions—small enough to be comprehended—and from which he receives large-scale answers too big for any human manager to have produced. Network management provides the algorithms, but human managers provide all the data and make all the decisions.

9

Waiting Lines and Service Times

INTRODUCTION

The managerial dilemma

When I was a young engineer, my duties included frequent sallies through the shops to see that no one was loafing. My training was technical, and no one had shown me the impossibility of arranging work flow in a multidepartment job shop so that men and machines were never idle. The mechanics adjusted skillfully to us beardless subalterns; it was quite a performance to see a veteran milling machine operator, his last job done and the next not yet arrived, contriving to look busy at an empty machine.

A little later my training program put me in a shipyard, helping a pipefitter named Duke install high-pressure piping in a submarine. Duke never worked steadily at anything: he'd template a pipe run and we'd go to the bench to assemble parts; then we'd be off to the bending floor to work pipe for a different system; then we'd go to the parts window and pick up some valves for a third job. He hopped back and forth between jobs, keeping five or six in the air all the time.

I thought Duke enjoyed variety, but it was just his way of beating waiting lines. He recognized that something has to give: if jobs are to move steadily ahead, excess people have to be standing by; if people are to be kept busy, spare jobs must be stockpiled.

Waiters and Servers

In any human interplay, there are people waiting for service, and there are people performing service. Duke was a *waiter;* his *server* was the man at the material window, the operator of the annealing furnace, or the draftsman preparing to modify an erroneous plan. Each was subject to unscheduled and variable demands from the mechanics, and not all demands could be satisfied without delay.

Customers are all too familiar with the external-internal queues where one side of the system falls under the manager's control and payroll, but the other does not. The doctor who makes appointments in advance, not knowing the complications of each patient, is torn between tight scheduling which keeps him busy and loose scheduling which keeps patients from waiting.* The doctor pays for delays on the serving side (by foregoing fees he could have if he were kept busy); but he cannot ignore the costs to the patients of inordinate delays on the waiting side, because ultimately he will pay a price for the latter. When the doctor calls a repairman to fix his broken x-ray machine, this is an external-internal queue in which he is the waiter. He pays the price directly (in missed business) of waiting for the equipment to be repaired; and indirectly he pays for idle time of the repairman between calls (since the repairman's rates must be high enough to compensate for such idle time). The person of moderate means patronizes a discount jeweler and waits in line to get reduced rates; the wealthy person pays more to an exclusive jeweler for being served at once.

It is not easy to estimate the price of delay to the external member of such a system. In a transportation study considering alternate transportation routes, a social valuation of $1.50 per hour was put on each driver's time; if a lower-cost route increased driving man-hours, the imputed cost of this wastage was included in the analysis. In commercial situations, the aim must be to convert the external waiting time into a cost to the serving business, but if this cannot be done there is some merit in pricing the cost to waiters (for ultimately they will move in the direction of action that will minimize these costs).

Managers of multiperson enterprises find internal-internal queues spread throughout their organization. In such queues, both waiter and server fall within the control and payroll of the organization, so that establishment of costs is direct. The mechanic who waits for material

---

* The exact looseness desired by patients is a moot point. Some studies suggest that patients want to be "where the action is," and will not continue to patronize a doctor whose waiting room is relatively empty.

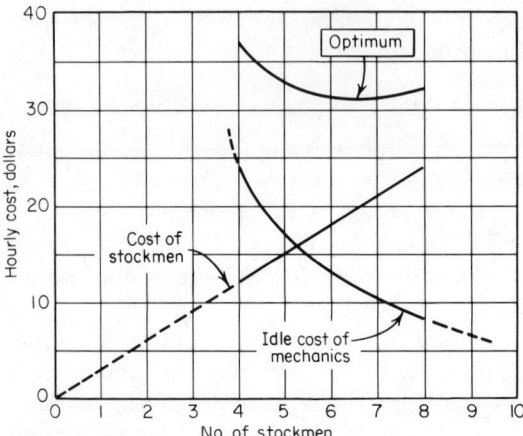

Fig. 9-1

at the shop store, the machine operator who waits for his machine to be repaired, the secretary who waits to use the duplicating machine, the truck driver who waits to get into the receiving bay—all these cost you their regular pay while in line.

Break-even Analysis

Suppose your problem is to minimize the cost of operating a tool issue room in an industrial shop, and you have been able to determine the total employee-hours spent in line each hour for various numbers of stockmen. Figure 9-1 shows the wage cost per hour for stockmen, each plotted against various numbers of stockmen. The sum of these two curves is the total personnel cost of the tool crib, and its minimum-cost point is the optimum number of stockmen. The curve indicates that 6 or 7 stockmen is the lowest-cost number.

This is very neat, but it leaves unanswered a most important question: How do you compute the idle-mechanic cost? (The cost of stockmen is quite obvious: simply their hourly wage, though this should be reduced by any backlog work you are able to schedule for their idle time.) This chapter suggests techniques for answering that question.

Queues are everywhere

The Dual Nature of Waiting Lines

The concept of waiters and servers is oversimplified, for actual situations cannot be cataloged so easily. A machine operator waiting for one of the repair gang to finish all outstanding jobs and get around to his disabled machine sees himself as the waiter; but a repairman sees himself as waiting his turn at a repair job, which "serves" him by bringing him work. The inventory situation is an even clearer example of duality. Imagine a single type of item in inventory and a single type of user. The items arrive in the bin one by one, on an irregular schedule, and the users call one by one in similar fashion. At times users get ahead of stock, and a queue of users forms awaiting "service"; at other times items get ahead of users, and a queue of items forms in the bin awaiting "service". There is a cost, inside the bin or outside it, for idleness; and since there will be idleness at times on both sides in any real case, the manager's problem is to apportion this idleness in the most economical way.

Examples of Waiting Lines

While the question of who is waiter and who is server may depend on the point of view, there isn't much confusion in most actual cases, such as the following:

A business telephone switchboard has outside lines that provide service when they are receiving calls. Customers who get busy signals constitute the waiting line. A switchboard position not in use generates unproductive cost; a customer not put through generates potential cost because he may be discouraged from patronizing the business.

A supermarket has a number of checkout booths, plus the problem of providing enough cashiers so that lines never grow too long and cashiers never stand idle too long.

An automobile dealer has mechanics on the payroll, and automobiles that arrive for repairs at unscheduled times. Cars awaiting repairs or customers waiting to be brought in are causes of customer dissatisfaction; mechanics awaiting jobs chew up profit.

A large metropolitan airport is a waiting-line nightmare at times, with airliners circling overhead, but if it is made big enough so that it scarcely ever has aerial traffic jams, it will have periods when its costly facilities lie fallow.*

A consulting firm must provide enough analysts to give fast service to customer firms, but must avoid costly staff sitting idle in the cubicles.

And the number of maintenance men in a hotel, the number of clerks in a store, the number of typists in a typing pool, the number of lanes in a superhighway, the number of loading stalls in a bus terminal, the number of students applying to a university for admission, the number of rooms in a motel—in short, any situation where the people or things to be serviced arrive in a variable way, and where the amount of service time or effort may be a variable too.

* A "textbook" waiting line is one whose arrivals are random, and airlines seek to avoid this by scheduling arrivals, but weather and mechanical problems often muddy up such plans to a point close to randomness.

Contents of this chapter

The purpose of this chapter is to help the manager find enough order in his waiting-line situation so that he can make the economical choice. This does not mean keeping service personnel or facilities occupied full time—which could make waiting lines grow without limit. Nor does it mean providing immediate service for all customers—which means gross excess in service facilities. It means recognizing the guideposts to an optimum course between these two infeasible extremes.

The chapter will show what starts queues and what makes them grow. It will show how to extract the inherent regularity from an apparently haphazard past experience and how to simulate various operating modes without disrupting actual operations. It will describe how to analyze waiting-line situations so that you can predict the important factors: average number in the line, average waiting time for any individual, rate of buildup of the line, and so on. It will show you how much improvement you get by increasing service capacity a specific amount.

There are truisms about waiting lines—things the manager should learn which are not intuitively obvious. You ought to have enough feel to appreciate these in a general way, even if you do no calculations. This chapter is designed to give you such a feel.

CHARACTERISTICS OF QUEUES

A desk-top waiting line

Place on the left side of your desk 5 nickels, and on the right side 3 pennies; toss the two piles separately every minute and record the two outcomes. The left pile is a gatekeeper: from the world of potential customers it makes one arrive every time a toss of 5 nickels comes up exactly 2 *heads*. The right pile decides how long it takes to service a customer: of the customers who have arrived and are waiting in line, it finishes servicing one every time a toss of 3 pennies comes up exactly 1 *head*. The probability of throwing 2 heads in a toss of 5 coins is 5 in 16 and the probability of throwing 1 head in a toss of 3 coins is 6 in 16; therefore, over the long run there will be 5 customers admitted every 16 minutes, and 6 customers serviced (if any are waiting) every 16 minutes.

This service ratio of 5:6 means you can take care of 6 customers in

WAITING LINES AND SERVICE TIMES   155

the time that 5 are arriving, *on the average.** Arrival times and service durations vary randomly about their average values, so this system will pulsate—at times there will be more people in line than the system can service, and at other times the servicemen will be idle. Since a serviceman cannot store up idle capacity as a squirrel stores up nuts, a minute idle is a minute lost forever, and this factor causes apparently favorable service ratios to create lines larger than you would expect.

You are using these coins to simulate a real situation whose arrival and service times behave somewhat as do the random tosses. On the average, you get 5 arrivals every 16 minutes, or an arrival every $3\frac{1}{5}$ minutes; and on the average you complete 6 services every 16 minutes, or a mean service rate of $2\frac{2}{3}$ minutes. If the actual situation you are trying to simulate has these *average times*, and if the *variation of individual times about the average* is distributed somewhat as coin-toss outcomes are distributed,† this desk-top simulation will provide useful information.

A 100-toss Simulation

If you were doing this in dead earnest, you would want many more than 100 simulations (and the computer would give you many more at very moderate cost), but even 100 tosses will be revealing. The results of such a series are shown in Table 9-1.

Follow Table 9-1 for a few tosses (each toss is taken as 1 minute of elapsed time):

Minute 1:   No arrivals (2 heads did not come up in left pile)
            No servicings (1 head did not come up in right pile)
            No customers in line
Minute 2:   Customer a arrives (2 heads came up)
            Customer a serviced (1 head came up)
            Customer a is in line and departs the same minute (a_0)
Minute 3:   Customer b arrives
            No servicings
            Customer b in line, does not depart (b_0)
Minute 4:   No arrivals or servicings
            Customer b stays in line, does not depart (b_1)

---

* A = arrival rate = 5/16 customer per minute, and S = servicing rate = 6/16 customer per minute. The service ratio is A/S = 5/6. (The term "service ratio" traditionally has different meaning than that applied here; it is used to express the long-run ratio between duration of service and duration of wait in line.)

† A binomial distribution, which means that the successive times are independent of one another, and that the long-run average arrival and service rates are constant.

Table 9-1
Arrival probability = [2H/5C] = $10/32$; arrival rate = $32/10$ = 3.2 minutes $\mu_A = 3\frac{1}{5}$
Service probability = [1H/3C] = $3/8$ = $12/32$; service rate = $32/12$ = 2.67 minutes $\mu_s = 2\frac{2}{3}$

Minute	Arrival	Service	Line	Minute	Arrival	Service	Line	Minute	Arrival	Service	Line
1	a	a	a_0	37	m	j	l_1 m_0 j_9 k_7	72	y	v	v_6 w_4 x_2
	b		b_0		n		l_2 m_1 j_{10} k_8 n_0			w	w_5 x_3
			b_1		o		l_3 m_2 o_0 k_9 n_1			x	x_4
5	c	b	b_2 c_0	40		k	l_4 m_3 o_1 k_{10} n_2				
	d	c	b_3 c_1			l	l_5 m_4 o_2 k_{11} n_3	75	z	y	y_0
			c_2 d_0				l_6 m_5 o_3 k_{12} n_4		a		y_1
			d_1	45	p	m	l_7 m_6 o_4 p_0 n_5		b		y_2
10	e	d	d_2		q	n	m_7 o_5 p_1 n_6				y_3 z_0
			d_3			o	q_0 m_8 o_6 p_2 n_7	80	c	z	y_4 z_1 a_0
			e_0				q_1 o_7 p_3 n_8				y_5 z_2 a_1 b_0
			e_1	50	r	p	q_2 o_8 p_4				z_3 a_2 b_1
15	f	e	e_2		s	q	q_3 p_5				z_4 a_3 b_2
			e_3 f_0			r	q_4 p_6	85	d	a	z_5 a_4 b_3
			e_4 f_1			s	q_5		e	b	c_0 a_5 b_4
			e_5 f_2	55	t	N_1	r_0				c_1 a_6 b_5
			e_6 f_3				s_0				c_2 a_7 b_6
20			e_7 f_4				t_0				c_3 a_8 b_7
			e_8 f_5			t	t_1	90		c	c_4 a_9 b_8
			f_6	60	u	u	t_2 u_0			d	c_5 d_0 a_{10} b_9 e_0
			f_7 g_0			N_2	t_3 u_1				c_6 d_1 a_{11} b_{10} e_1
25	g	f	f_8 g_1				u_2				c_7 d_2 a_{12} b_{11} e_2
	h	g	f_9 g_2				u_3				c_8 d_3 a_{13} b_{12} e_3
		h	g_3	65		N_3		95	f		c_9 d_4 a_{14} b_{13} e_4
									g		c_{10} d_5 a_{15} b_{14} e_5
30	i		h_0								c_{11} d_6 a_{16} b_{15} e_6
	j		h_1	70	v	N_4	v_0				c_{12} d_7 f_0 b_{16} e_7
	k		h_2 i_0		w		v_1				c_{13} d_8 f_1 g_0 e_8
			i_1 j_0		x		v_2 w_0				c_{14} d_9 f_2 g_1 e_9
			i_2 j_1 k_0				v_3 w_1	100	h	e	c_{15} d_{10} f_2 g_2 e_{10}
35	l	i	i_3 j_2 k_1				v_4 w_2 x_0		i		h_0 d_{11} f_3 g_2 e_{11}
			i_4 j_3 k_2				v_5 w_3 x_1		j		h_1 i_0 f_4 g_3 e_{12} j_0
			i_5 j_4 k_3								h_2 i_1 f_5 g_4
			i_6 j_5 k_4								
			i_7 j_6 k_5								
			i_8 j_7 k_6								
			l_0 j_8								

Minute 5:   Customer c arrives
No servicings
Two customers in line; b for his third minute, c for his first minute (b_2, c_0)

And so on. Now skip down to minute 54. The waiting line has dwindled to zero, as servicings have caught up with arrivals, and the right coin toss indicates another servicing—but with no customers, no servicing can be done. This is a wasted service minute, noted as N_1 or the first such wasted service unit.

Simulation Results

What were the results of this simulation?

Average size of waiting line: Counting the number of times there are no customers in line, 1 customer in line and so on, and averaging, give the results shown in Table 9-2.

Average time spent in line: You can compute the average time in line by tabulating the number of occasions when 1 minute is spent, the number when 2 minutes are spent, and so on, as you did in Table 9-2. Such a calculation is shown in Table 9-3.

Maximum length of line: Table 9-2 shows a maximum of 6 people in line at any time, and that only once in 100 times; but the line was 5 or longer 20 percent of the time. You are interested less in the averages than you are some upper limit which contains a significant minority of the cases. Perhaps you can ignore a situation that happens 1 in 100 times, but if it happens 1 in 5 times your customers will begin to feel it is typical.

Table 9-2

Length of waiting line, L	Frequency with which this length occurs, F	Product, $L \times F$
0	9	0
1	22	22
2	25	50
3	18	54
4	6	24
5	19	95
6	1	6
	100	251

$L_{\text{average}} = 251/100 = 2.51$ in line (counting the one being served)

Table 9-3

Time spent in line (including servicing), T	Frequency of customers spending this time, F	Product, F × T
0	3	0
1	0	0
2	2	4
3	5	15
4	1	4
5	4	20
6	2	12
7	1	7
8	5	40
9	1	9
10	1	10
11	1	11
12	2	24
13	0	0
14	0	0
15	1	15
16	2	32
	31	203

$T_{\text{average}} = {}^{203}\!/_{31} = 6.55$ minutes

Maximum length of time in line: The customer is disturbed by the length of a line, but more significant is the amount of time he must wait. Table 9-3 shows that a customer can expect to wait 16 minutes 2 times out of 31, and can expect to wait 11 *minutes or more* 20 *percent of the time*. The 6½-minute average wait is less important than the wait of 11 minutes or more that occurs to a significant minority.

Buildup of the Line

These figures do not tell the real story, because this 100-minute simulation occurred before the system had a chance to reach steady-state conditions. A calculation similar to that of Table 9-2 for the first 20 minutes, the second 20 minutes, and so on, would look like Table 9-4. When the simulation started, there was no waiting line, so the first 20 minutes was not representative. Table 9-4 suggests that there is a buildup of the waiting line as time goes on, to a point where it reaches a steady state (in which the length of the line varies up and down, but with no more underlying upward trend). As the service ratio gets closer to unity, the steady-state length approaches infinity.

WAITING LINES AND SERVICE TIMES

Table 9-4

No. in line, L	First 20 minutes		Second 20 minutes		Third 20 minutes		Fourth 20 minutes		Fifth 20 minutes	
	F	$F \times L$	F	$F \times L$	F	$F \times L$	F	$F \times L$	F	$F \times L$
0	2	0	—	0	2	0	5	0	—	0
1	9	9	2	2	7	7	4	4	—	0
2	9	18	7	14	4	8	5	10	—	0
3	—	0	7	21	1	3	5	15	5	5
4	—	0	1	4	2	8	1	4	2	8
5	—	0	3	15	4	20	—	0	12	60
6	—	0	—	0	—	0	—	0	1	6
		27		56		46		33		89
L_{average}		1.35		2.8		2.3		1.65		4.45

This suggests that some waiting lines, whose steady-state length could be very high, are saved by the intermittent nature of one-shift use. Hospital operating rooms may be scheduled routinely from 8 A.M. to 4 P.M., but the buildup of patients awaiting operations will be drawn off every evening by using operating rooms past 4 P.M. whenever there is a line. Steady-state conditions might be completely unacceptable, but since it takes more than eight hours to reach them (with the average operation taking perhaps an hour), they never occur. The hospital would be in for a rude shock if it were to institute 24-hour scheduling of operating rooms (converting two-thirds of the operating rooms to other uses), for it would move into a steady-state mode and vastly changed circumstances.

Validity of This Simulation

Granting that this simulation was too brief, how close were the coin-toss outcomes to the expected values? Probability theory says that, in the long run, 2 heads in 5 coins will come up 31.3 percent of the time, and 1 head in 3 coins will come up 37.5 percent of the time. Actual simulation outcomes were 36 arrivals and 35 services (4 of which were unused) out of 100. These variations are not out of line.

Table 9.5 carries this simulation 100 minutes further, using the same coin-toss outcomes as the first 100, but starting with waiting-line conditions of the one-hundredth minute. The results of this show how far the first simulation was from steady state: the average wait for the second 100 minutes of Table 9-5 is 15.6 minutes, with an average of 6 in the line.

Table 9-5

Min.	Arr.	Svc.	Line
101			$f_6\ g_5\ h_3\ i_2\ j_1$
2	k	f	$f_7\ g_6\ h_4\ i_3\ j_2\ k_0$
3	l		$g_7\ h_5\ i_4\ j_3\ k_1\ l_0$
4			$g_8\ h_6\ i_5\ j_4\ k_2\ l_1$
5	m		$g_9\ h_7\ i_6\ j_5\ k_3\ l_2\ m_0$
6		g	$g_{10}\ h_8\ i_7\ j_6\ k_4\ l_3\ m_1$
7	n	h	$h_9\ i_8\ j_7\ k_5\ l_4\ m_2\ n_0$
8			$i_9\ j_8\ k_6\ l_5\ m_3\ n_1$
9			$i_{10}\ j_9\ k_7\ l_6\ m_4\ n_2$
10		i	$i_{11}\ j_{10}\ k_8\ l_7\ m_5\ n_3$
11			$j_{11}\ k_9\ l_8\ m_6\ n_4$
12	o		$j_{12}\ k_{10}\ l_9\ m_7\ n_5\ o_0$
13			$j_{13}\ k_{11}\ l_{10}\ m_8\ n_6\ o_1$
14			$j_{14}\ k_{12}\ l_{11}\ m_9\ n_7\ o_2$
15	p		$j_{15}\ k_{13}\ l_{12}\ m_{10}\ n_8\ o_3\ p_0$
16			$j_{16}\ k_{14}\ l_{13}\ m_{11}\ n_9\ o_4\ p_1$
17			$j_{17}\ k_{15}\ l_{14}\ m_{12}\ n_{10}\ o_5\ p_2$
18			$j_{18}\ k_{16}\ l_{15}\ m_{13}\ n_{11}\ o_6\ p_3$
19			$j_{19}\ k_{17}\ l_{16}\ m_{14}\ n_{12}\ o_7\ p_4$
20		j	$j_{20}\ k_{18}\ l_{17}\ m_{15}\ n_{13}\ o_8\ p_5$
21			$k_{19}\ l_{18}\ m_{16}\ n_{13}\ o_9\ p_6$
22	q		$k_{20}\ l_{19}\ m_{17}\ n_{14}\ o_{10}\ p_7\ q_0$
23			$k_{21}\ l_{20}\ m_{18}\ n_{15}\ o_{11}\ p_8\ q_1$
24		k	$k_{22}\ l_{21}\ m_{19}\ n_{16}\ o_{12}\ p_9\ q_2$
25	r	l	$l_{22}\ m_{20}\ n_{17}\ o_{13}\ p_{10}\ q_3\ r_0$
26			$m_{21}\ n_{18}\ o_{14}\ p_{11}\ q_4\ r_1$
27	s	m	$m_{22}\ n_{19}\ o_{15}\ p_{12}\ q_5\ r_2\ s_0$
28	t		$n_{20}\ o_{16}\ p_{13}\ q_6\ r_3\ s_1\ t_0$
29			$n_{21}\ o_{17}\ p_{14}\ q_7\ r_4\ s_2\ t_1$
30	u		$n_{22}\ o_{18}\ p_{15}\ q_8\ r_5\ s_3\ t_2\ u_0$
31			$n_{23}\ o_{19}\ p_{16}\ q_9\ r_6\ s_4\ t_3\ u_1$
32			$n_{24}\ o_{20}\ p_{17}\ q_{10}\ r_7\ s_5\ t_4\ u_2$
33			$n_{25}\ o_{21}\ p_{18}\ q_{11}\ r_8\ s_6\ t_5\ u_3$
34			$n_{26}\ o_{22}\ p_{19}\ q_{12}\ r_9\ s_7\ t_6\ u_4$
35		n	$n_{27}\ o_{23}\ p_{20}\ q_{13}\ r_{10}\ s_8\ t_7\ u_5$
36	v		$o_{24}\ p_{21}\ q_{14}\ r_{11}\ s_9\ t_8\ u_6\ v_0$
37	w		$o_{25}\ p_{22}\ q_{15}\ r_{12}\ s_{10}\ t_9\ u_7\ v_1\ w_0$
38	x	o	$o_{26}\ p_{23}\ q_{16}\ r_{13}\ s_{11}\ t_{10}\ u_8\ v_2\ w_1\ x_0$
39	y		$p_{24}\ q_{17}\ r_{14}\ s_{12}\ t_{11}\ u_9\ v_3\ w_2\ x_1\ y_0$
40			$p_{25}\ q_{18}\ r_{15}\ s_{13}\ t_{12}\ u_{10}\ v_4\ w_3\ x_2\ y_1$
41			$p_{26}\ q_{19}\ r_{16}\ s_{14}\ t_{13}\ u_{11}\ v_5\ w_4\ x_3\ y_2$
42		p	$p_{27}\ q_{20}\ r_{17}\ s_{15}\ t_{14}\ u_{12}\ v_6\ w_5\ x_4\ y_2$
43	z	q	$q_{21}\ r_{18}\ s_{16}\ t_{15}\ u_{13}\ v_7\ w_6\ x_5\ y_3\ z_0$
44			$r_{19}\ s_{17}\ t_{16}\ u_{14}\ v_8\ w_7\ x_6\ y_4\ z_1$
45	a	r	$r_{20}\ s_{18}\ t_{17}\ u_{15}\ v_9\ w_8\ x_7\ y_5\ z_2\ a_0$
46		s	$s_{19}\ t_{18}\ u_{16}\ v_{10}\ w_9\ x_8\ y_6\ z_3\ a_1$

WAITING LINES AND SERVICE TIMES

Table 9-5 (continued)

Min.	Arr.	Svc.	Line
47		t	$t_{19}\ u_{17}\ v_{11}\ w_{10}\ x_9\ y_7\ z_4\ a_2$
48			$u_{18}\ v_{12}\ w_{11}\ x_{10}\ y_8\ z_5\ a_3$
49		u	$u_{19}\ v_{13}\ w_{12}\ x_{11}\ y_9\ z_6\ a_4$
50		v	$v_{14}\ w_{13}\ x_{12}\ y_{10}\ z_7\ a_5$
51	b	w	$w_{14}\ x_{13}\ y_{11}\ z_8\ a_6\ b_0$
52	c	x	$x_{14}\ y_{12}\ z_9\ a_7\ b_1\ c_0$
53			$y_{13}\ z_{10}\ a_8\ b_2\ c_1$
54		y	$y_{14}\ z_{11}\ a_9\ b_3\ c_2$
55	d		$z_{12}\ a_{10}\ b_4\ c_3\ d_0$
56			$z_{13}\ a_{11}\ b_5\ c_4\ d_1$
57	e		$z_{14}\ a_{12}\ b_6\ c_5\ d_2\ e_0$
58		z	$z_{15}\ a_{13}\ b_7\ c_6\ d_3\ e_1$
59			$a_{14}\ b_8\ c_7\ d_4\ e_2$
60		a	$a_{15}\ b_9\ c_8\ d_5\ e_3$
61		b	$b_{10}\ c_9\ d_6\ e_4$
62			$c_{10}\ d_7\ e_5$
63		c	$c_{11}\ d_8\ e_6$
64			$d_9\ e_7$
65		d	$d_{10}\ e_8$
66	f		$e_9\ f_0$
67			$e_{10}\ f_1$
68	g		$e_{11}\ f_2\ g_0$
69			$e_{12}\ f_3\ g_1$
70	h		$e_{13}\ f_4\ g_2\ h_0$
71			$e_{14}\ f_5\ g_3\ h_1$
72		e	$e_{15}\ f_6\ g_4\ h_2$
73		f	$f_7\ g_5\ h_3$
74	i	g	$g_6\ h_4\ i_0$
75			$h_5\ i_1$
76			$h_6\ i_2$
77	j		$h_7\ i_3\ j_0$
78	k		$h_8\ i_4\ j_1\ k_0$
79	l	h	$h_9\ i_5\ j_2\ k_1\ l_0$
80			$i_6\ j_3\ k_2\ l_1$
81			$i_7\ j_4\ k_3\ l_2$
82	m	i	$i_8\ j_5\ k_4\ l_3\ m_0$
83			$j_6\ k_5\ l_4\ m_1$
84			$j_7\ k_6\ l_5\ m_2$
85			$j_8\ k_7\ l_6\ m_3$
86			$j_9\ k_8\ l_7\ m_4$
87	n		$j_{10}\ k_9\ l_8\ m_5\ n_0$
88	o		$j_{11}\ k_{10}\ l_9\ m_6\ n_1\ o_0$
89			$j_{12}\ k_{11}\ l_{10}\ m_7\ n_2\ o_1$
90			$j_{13}\ k_{12}\ l_{11}\ m_8\ n_3\ o_2$
91			$j_{14}\ k_{13}\ l_{12}\ m_9\ n_4\ o_3$
92			$j_{15}\ k_{14}\ l_{13}\ m_{10}\ n_5\ o_4$

Table 9-5 (continued)

Min.	Arr.	Svc.	Line
93			$j_{16}\ k_{15}\ l_{14}\ m_{11}\ n_6\ o_5$
94		j	$j_{17}\ k_{16}\ l_{15}\ m_{12}\ n_7\ o_6$
95	p	k	$k_{17}\ l_{16}\ m_{13}\ n_8\ o_7\ p_0$
96	q		$l_{17}\ m_{14}\ n_9\ o_8\ p_1\ q_0$
97		l	$l_{18}\ m_{15}\ n_{10}\ o_9\ p_2\ q_1$
98	r	m	$m_{16}\ n_{11}\ o_{10}\ p_3\ q_2\ r_0$
99	s		$n_{12}\ o_{11}\ p_4\ q_3\ r_1\ s_0$
200	t	n	$n_{13}\ o_{12}\ p_5\ q_4\ r_2\ s_1\ t_0$

$$T_{\text{aver}} = \frac{1}{1-\rho} \times \text{service time}$$

$$6 \times 2.67 = 16$$

$$T_{\text{aver}} = \frac{546}{35} = 15.6$$

$$L_{\text{aver}} = \frac{\rho}{1-\rho} = \frac{5/6}{1/6} = 5$$

$$L_{\text{aver}} = \frac{603}{100} = 6$$

L	F	L×F
2	6	12
3	7	31
4	12	48
5	10	50
6	28	168
7	19	133
8	8	64
9	3	27
10	7	70
		603

Why queues start

Fixed Arrival and Service Times

Imagine a situation similar to the desk-top simulation, but with arrival and service times fixed at their average values. Every $3\frac{1}{5}$ minutes a customer arrives, and every servicing takes exactly $2\frac{2}{3}$ minutes. A queue cannot form at all, because the serviceman always will finish with one customer before the next one arrives. Raise the service ratio to unity; every servicing now takes exactly $3\frac{1}{5}$ minutes, and a queue still cannot form, because the serviceman will finish with each customer the instant the next arrives. Raise the service ratio to a bit above unity—say the serviceman takes $3\frac{2}{5}$ minutes a customer—and the line starts to grow; but at a regular and rather modest rate of $\frac{1}{5}$ minute per customer. It would take 16 customers before the line had increased one person, and if the line stopped every 20 customers or so to catch up as the operating rooms did, this would be tolerable.

If waiting lines met these happy conditions, theory would be a waste of time; but despite efforts to schedule to such an ideal, they seldom do.

WAITING LINES AND SERVICE TIMES

Random Arrival and Fixed Service Times

As soon as one element of the pair varies in a random way, the potential for generation of a waiting line exists. Suppose you have a gasoline station on a turnpike, with servicing time of exactly 50 seconds; a car passes every 30 seconds, and half the cars on the average stop for gasoline. The service ratio is the same as for the desk-top simulation, in which both elements varied randomly. Simulate the situation by tossing a coin every 30 seconds, to decide whether the car comes in or goes by; Fig. 9-2 shows 50 trials of such a simulation.

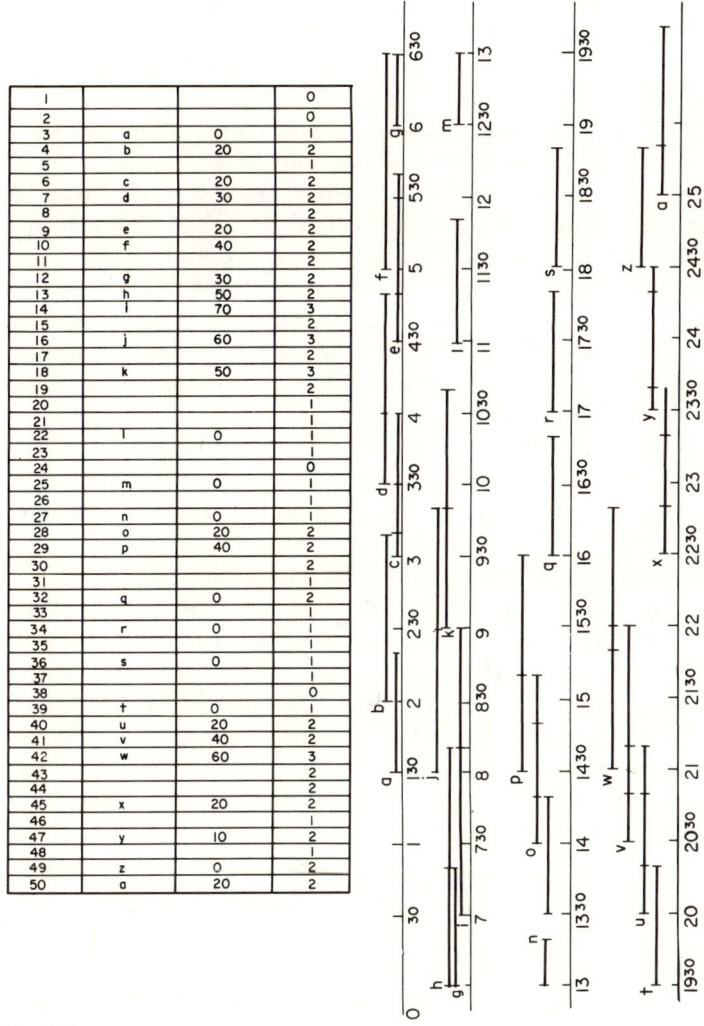

Fig. 9-2

This simulation shows an average of 1.56 cars in line, counting the one being serviced. There is an average wait of 22.6 seconds before servicing starts, and nearly 20 percent of the time there is a wait of 50 seconds or more.

Random Arrival and Random Service Times

How does the above situation, with one of the two elements fixed, compare with the desk-top situation, where both elements varied randomly? It approaches steady state faster, and the waiting line is shorter for the same service ratio. Most waiting situations have random elements in both arrival and service times; and since these situations produce the most troublesome lines, queueing theory devotes most attention to the case where both vary randomly.

Effect of Service Ratio

Probability of delay: The service ratio (A/S) is a rough approximation, in many such waiting-line situations, of the probability that a customer will encounter some delay. In the desk-top case, the service ratio is 5:6 or 83 percent. Table 9-1 shows that 29 of 36 arrivals (80 percent) found the preceding customer still in line or being serviced.*

The square of this ratio approximates the probability that the line will have two or more people in it: $(5:6)^2$ or 69 percent in this case, compared with 52 percent for the coin-toss simulation. The probability that the line will have n or more people in it when a customer arrives is approximated by $(A/S)^n$. The calculated probability of four or more people in line ahead of a customer is $(5:6)^4$ or 48 percent; the desk-top simulation had 28 percent. It must be remembered, however, that these formulas are intended to apply to steady-state conditions, which were not attained in 100 simulations.

You can see the impossible situation this relationship imposes on you when you attempt to provide enough service so there will be no waiting. If by "no waiting" you mean a customer waits 1 time in 10, your customer service organization will be only 10 percent utilized.† If the cost of service is cheap compared with the cost of customer delay—such as a jewelry store catering to clients so wealthy that every customer brings in large commissions—you will accept service ratios close to zero; but you need an appreciation of the theory so that you can decide just where your optimum strategy lies.

* This approximation is more accurate with certain waiting-line models which represent actual waiting-line situations better than does the coin-toss.

† You can have standby work for the service group to do when there are no customers, but if this is a feasible solution you have no need of waiting-line theory in the first place.

WAITING LINES AND SERVICE TIMES

Length of line: The previous section points out that the probability of *some* delay rises approximately in proportion to service ratio: if you double the latter, you tend to double the former. This is a linear relationship, pretty well behaved. But the *amount* of delay, as service ratio increases, isn't so well behaved. When service ratio for a system with random arrival and service times gets within shouting distance of 100 percent, line length shoots through the roof.

After such a system reaches its steady state, the average waiting time for a customer is given rather closely by the formula:

$$T_{\text{average}} = \left(\frac{1}{1 - \text{service ratio}}\right) \times \text{mean service rate}$$

If the service ratio is 5:6, as in the desk-top simulation, this works out to $1/(1 - 5/6) \times 2.67 = 16$.

If the service ratio had been 9:10, the waiting time would have been 27 minutes, and if it had been 99:100, the waiting time would have been 267 minutes. Figure 9-3 shows how the value $1/(1 - \text{service ratio})$ increases exponentially with increase in service ratio.

The average length of line in the second 100 desk-top simulations is 15.6—quite close to the value of 16 given by the formula above.

Number in line: The average number in line after such a system reaches steady state is given approximately by the formula $L_{\text{average}} = \text{service ratio}/(1 - \text{service ratio})$. For our ratio it is $5:6/1:6 = 5$. The

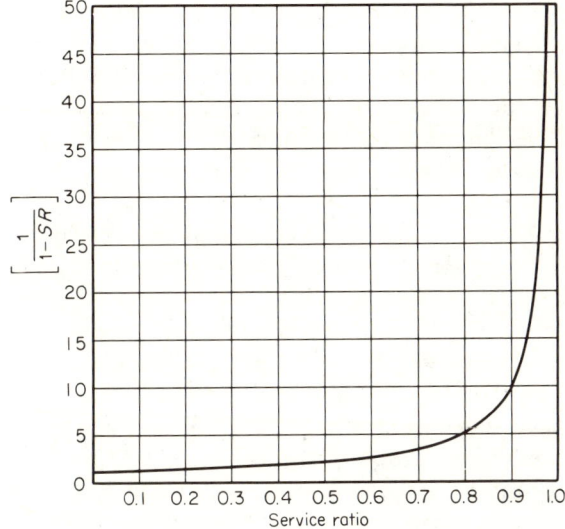

Fig. 9-3

length of line for a service ratio of 9:10 would be 9, and for a service ratio of 99:100 it would be 99.

The average length of line for the second 100 desk-top simulations is 6.0—reasonably close to the value given by the formula.

Why queues grow

The difference between the first 100 and the second 100 desk-top simulations suggests why waiting lines grow. When you started with toss 1, the line was empty, and the first "customer" was served immediately. By toss 101 the line had built up to 6 awaiting or receiving service, and the second 100 started behind the eight ball. A third 100 simulations would start with 7 in line; since this is not very different from the 6 who faced the second 100, you suspect that you are nearing steady state.

What was different between the first 100 and second 100 that induced growth in the former but caused the latter to taper off? If the line grows for a time, why does it ever stop? The answer lies in the missed opportunities of the growth phase—the service potential at minutes 54, 61, 63, and 65 that was wasted because no customers were in line. The second 100 had no such waste, so the servicing mechanism was working to full effectiveness.

The 200-minute simulation had 72 arrivals, and the buildup did not occur until nearly half this number had arrived. Figure 9-4 shows a case with a service ratio of 96 percent: the buildup does not start until 100 arrivals, but at 250 arrivals it is still rising. When traffic builds up to such delays, it will be extremely slow to disperse unless really drastic efforts are made; the service ratio must be lowered markedly below what is acceptable for steady state in order to bring the line down in a hurry. Figure 9-3 shows a marked upturn in the curve at 80 percent service ratio. This ratio of 80 percent is a ceiling that you should not try to breach, until you have made detailed calculation or simulation of your particular situation.

Note one significant factor about waiting lines that may rescue an overloaded situation. If you never reach steady state, it hardly matters how bad that state can be. In the desk-top simulation, if your 8 A.M. to 5 P.M. daily routine never exceeds the first-100 condition, the second-100 congestion is just theory. In Fig. 9-4, even a 96 percent ratio is acceptable if the daily cycle stays under 100 arrivals. Analytical solutions apply generally to conditions in a steady state that you may never reach. Peak-hour airline traffic conditions may tax the airfield capacity sorely, and calculation may show intolerable waiting lines—but one hour isn't long enough for this to happen, and a light hour following may save the situation.

WAITING LINES AND SERVICE TIMES

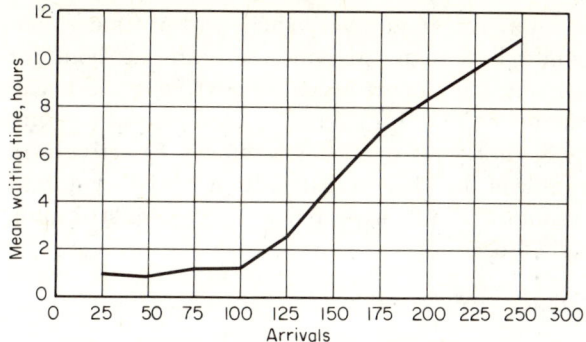

Fig. 9-4 (*From E. Duckworth, A Guide to Operational Research,* Methune & Co., Ltd., *University Paperbacks, London,* 1965.)

Simulation sheds light on these transient conditions when standard mathematical formulas do not. You have seen this work with coin-toss simulations. In actual waiting lines, the coin toss may not be the best imitation of reality, and you will require different methods. These will be discussed a bit in the next section.

ANALYSIS OF QUEUES

Mathematical distributions

Your operations researcher will tell you that, in real life, arrivals which seem haphazard are likely to follow a "Poisson" distribution and service times are likely to be distributed in an "exponential" way. What are these two distributions?

The Poisson Distribution

This distribution is a relative of the coin toss, but with a difference. In the desk-top experiment, you tossed a coin every minute and it came up "yes" or "no"—there was no arrival or 1 arrival. In real life there might have been 2 arrivals in that minute, or 3, or 10. All right, toss a coin every second; but if you want to keep the same probability of 1 arrival per minute you'll need a 120-sided coin with 1 head and 119 tails—which will give an expectation of $\frac{1}{120} \times 60$ or $\frac{1}{2}$ head per minute. But there might be 2 arrivals in a single second; you therefore subdivide further by tossing a 1,200-sided coin with 1 head 600 times a minute. If you continue this subdividing down to virtually no time at all, with an infinite-sided coin, the coin toss merges into a distribution called "Poisson."

The Poisson distribution simulates many real-life arrival situations where the probability of an arrival in any very small unit of time is constant, and where you know how many arrivals there are per hour (or other unit) *on the average* but don't know what the next hour will bring. The Poisson is completely defined by this average arrival rate (A). Suppose your records tell you that your operation averages 10 arrivals an hour. If you consult a table of the Poisson distribution,* it tells you that 100 hours selected at random would be expected to have the distribution of arrivals shown in Table 9-6.

The Exponential Distribution

This distribution often characterizes the variation of service times. If your average service rate is S ($6/16$ customer per minute in the desk-top simulation), the average service time is $1/S$ ($2\frac{2}{3}$ minutes per customer in the desk-top case). The probability that a servicing will be completed in 1 minute is S ($6/16$ in the desk-top case), another way of expressing the fact that in the long run 6 servicings are completed every 16 minutes.

If you start miniaturizing this relationship as you did the arrival situation, the probability of completing servicing of a waiting customer in $\frac{1}{2}$ minute is $\frac{1}{2}S$, in $\frac{1}{10}$ minute is $\frac{1}{10}S$, and so on down to the infini-

* Such as E. C. Molina, *Poisson's Exponential Binomial Limit*, D. Van Nostrand Company, Inc., Princeton, N.J., 1942.

Table 9-6

Number of arrivals per hour	Expected frequency of this number per 100 random arrivals
3 or less	1
4	2
5	4
6	6
7	9
8	11
9	13
10	13
11	11
12	10
13	7
14	5
15	3
16	2
17	1
18 or more	2

WAITING LINES AND SERVICE TIMES

tesimal point where it becomes a mathematical distribution known as the exponential.

There is a fundamental difference between the Poisson and exponential distributions: arrivals come in whole numbers (you can't have $4\frac{1}{2}$ customers) but service time can be any fractional value. This leads to the minor nuisance that, since there are infinitely many times, it is meaningless to express the probability of each; but you can give the probability that a time is less than some stated time, or is between some time limits. Table 9-7 shows an exponential distribution of service times when the average service time $(1/S)$ is 10 minutes.

The probability that a servicing will take between 9 and 10 minutes is 63 percent − 59 percent or 4 percent, the probability that it will take between 1 and 2 minutes is 9 percent, and so on.

Mathematical formulas

Some basic waiting-line formulas are given above. There are formulas for calculating various other values of waiting lines: average length of nonempty lines, probability of waiting more than any specified time, ratio of time served to time waiting, and so on. These calculations vary with the type of waiting line: more than one serving station, special priority rules for being served, rules for customers leaving unsatisfied after specified times, and so on. There are compound waiting lines: the sequence of lines involved in successive manufacturing operations, the transmission of sudden increase in market demand back to the factory and up to the customer in the form of increased production, and so on. Such pipeline situations often produce increasing shock waves such as those caused by sudden stops on crowded turnpikes, with the final result far out of proportion to the cause. The mathematics of such waiting-line sequences is quite complex.

A drawback to many mathematical formulas is that they apply after the system reaches steady state, and you have seen that this may not occur for a long time (or, in cases where there are regular reductions in traffic, may not occur at all). Simulation of a waiting-line system will show its transient behavior—how long before it reaches steady state, and the way it behaves en route—and simulation requires no more than a knowledge of the arrival and service distributions and the use of a random number table.

Simulation

You used a coin to simulate arrival and service times, which is fine if the actual events follow such a pattern. Frequently they do not, so you

Table 9-7

Service time (t), minutes	Probability that a random servicing takes less than this service time, %*
1	9
2	18
3	26
4	33
5	39
6	45
7	50
8	55
9	59
10	63
11	67
12	70
13	73
14	75.5
15	78
16	80
17	82
18	83.5
19	85
20	86.5
21	88
22	89
23	90
24	91
25	92
30	95
35	97
40	98
45	99
50	99.5
60	100 (nearly)

* These probabilities can be computed by substituting the various values of service time into the formula: Probability $= 1 - e^{-st}$, where e is the base of the natural logarithmic system (2.71828 . . .), and s is the average service rate.

must investigate to see whether they do or don't. If you conclude, by analyzing the factors that cause arrival and service times to vary, that the conditions for Poisson and exponential distributions exist (if the probability that a customer arrives, or the probability that a servicing is completed, is constant for each small period of time), your situation is described with reasonable accuracy by these distributions. If not, you must take a rather large number of observations of your operation over a representative period, and convert them into frequency tables such as Tables 9-6 and 9-7. When you have done one of these two things, you are ready to run a simulation.

Random Number Simulation

You will use a random number table as the equivalent of a 100-sided die. Many statistics books and all computers contain random number tables, which are simply lists of numbers arranged in such random order that if you go down the list from some random starting point, every number from 0 to 99 (or to 999, or 9,999, or whatever number of digits the table provides) has a theoretically equal chance of being encountered next. Since you are interested only in a two-digit set, ignore any but the first two digits.

Suppose you are running a simulation based on Table 9-6. Dive into the table at random and note the first number. If it is 00, let this translate into "3 or fewer arrivals per hour"; if 01 or 02, "4 arrivals per hour"; if 03 through 06, "5 arrivals per hour"; and so on down Table 9-6. This is your first simulation. Do this 100 times, noting the corresponding number of arrivals for each of the 100 simulated hours, and you have run a 100-event simulation of arrival times. Do the same thing all over again for service times. Now you can convert these into a waiting-line table such as Table 9-1, and examine it in detail to see how it is behaving. You will be interested in finding the size of the line, as in Table 9-2; the time spent in line, as in Table 9-3; and how fast it is approaching steady state (by starting with the line that exists after the one-hundredth simulation and using the same arrival and service times to run a second 100).

You are not finished. You have played the simulation game once, and have learned some interesting things, but it won't always come out this way. Play the game several times, and see how much variation you get in average line size and average time spent in line. After 10 or 12 such plays, the results should be as accurate as you need them, but if there is much variation in the averages from each play you can continue until you get consistent grouping of results.

This sounds like a lot of work, but it is far cheaper to simulate than to go through actual operations. (A computer runs simulations fast and economically.) The major contribution of simulation is its ability to tell you what happens if you change a key element, *without* trying such a change in actual operations and perhaps throwing your business into chaos. You can see what happens if you change the average service time ($1/S$), or if arrivals speed up by some specific amount. You can see the effect of adding or taking away servicemen. You can live through situations vicariously, without pain or strain.

GUIDELINES AND CONCLUSIONS

By now you have reached several general conclusions. In any situation where arrival and service times are haphazard, it is fruitless to strive for a service facility where customers never have to wait and servicemen are never idle. Once you get above a service ratio of about 80 percent, be prepared for growing queues. Once a line has built up to substantial length, it will take a drastic reduction in service ratio to bring it down fast. If you can shut your line off at regular intervals (such as by going home at night), you may be able to get away with extremely high service ratios—but you must simulate to be sure. Average waits are important, but you must check the longer wait which a significant majority of customers will encounter—say the top 10 or 15 percent. Processes involving sequences of waiting lines can magnify small variations tremendously—a temporary 10 percent increase in retail market demand can trigger an unwanted 40 percent increase in production.

And if you want to make a change in your operations, you don't need to guess at the effects—simulate!

10

Programming

WHAT PROGRAMMING DOES

The need for programming

When all management problems are simple and the way clearly marked, there is no need for programming or any other quantitative aid. If your plant makes only one product, your formulation has only one possible mix of ingredients, your machines do only one job, your delivery service takes only one possible route, your funds face only one possible investment opportunity—then you have no problems of what to put where. Even if your situation is a bit more complex, experience or trial and error may find nearly the best allocation of resources; if you are assigned 3 jobs and dispatching 3 trucks, it isn't hard to consider each of the 6 different alternatives. But complexities pyramid swiftly: make that 10 jobs and 10 trucks, and now you must consider $3\frac{1}{2}$ million alternatives.

The impossibility of coping rationally with such choices by unaided managerial judgment has led to the development of quantitative techniques for allocating resources to tasks in the best way. Mathematicians call such techniques "powerful," and so they are in the sense that they find their way through muddles of numbers and come up with the best answer *for those numbers*. But they are no more powerful than the accuracy of the numbers management provides. A significant side benefit of programming is the fact that it requires managers to make numerical judgments about matters they have been treating subjectively, and to look to the accuracy of their decision-making data.

The role of programming

This chapter tackles the problem of assigning resources to tasks where there are limitations on the resources, and varying payoffs on the tasks, in such a way as to optimize the results—maximize output for a given input, minimize input for a given output, or generally produce the largest margin of return in relation to the cost. Three principal techniques will be discussed: the assignment method, to meet problems such as optimal assignment of jobs to machines; the transportation method, to solve tasks such as optimum routing in transportation networks; and the linear programming method, to deal with more general questions of resource allocation.

THE ASSIGNMENT PROBLEM

Application

Suppose you have three trucks and three types of deliveries to be made. The trucks differ in size and the three types of material are different, so that for a given delivery there is a different cost for using each of the three trucks, as shown in Table 10-1.

Step-by-step solution

Loss Table for Equipment Choice

If you consider each delivery separately, it is easy to determine which truck is the most economical choice—and you compute for each of the

Table 10-1

	Truck 1	Truck 2	Truck 3
Delivery type A	$55	$60	$70
Delivery type B	$60	$75	$80
Delivery type C	$70	$85	$95

Table 10-2

	Equipment 1	Equipment 2	Equipment 3
Job A	0	5	15
Job B	0	15	20
Job C	0	15	25

other two trucks the "loss" you would incur by using it instead of the optimum one. For job A, truck 1 would be optimum; you would have a loss of $5 if you used truck 2 and a loss of $15 if you used truck 3. Computing the loss for each of the other two jobs in the same way would give the results shown in Table 10-2.

Loss Table for Job Choice

Suppose you did it the other way: decided for each truck which job it could do best (at zero "loss"), and what relative loss you would incur by employing it for jobs other than this optimum job. For truck 1, the job-choice losses would be:

Job A $ 0
Job B   5
Job C   15

The complete job-choice loss table would be as shown in Table 10-3.

Combined Loss Table

Each of the above tables leaves something to be desired. In the equipment-choice table, truck 1 is clearly best for each job—but since you can't assign more than one job to a truck, this isn't much help. In the job-choice table, each truck rates job A as the preferred choice, but only one truck can get job A. Clearly this problem can't be solved by either of these simple approaches. Try a first approximation to a combined approach. Make an equipment-choice loss table as you did above, and then make a job-choice table *from that*. A job-choice loss table made from Table 10-2 would look as shown in Table 10-4.

Table 10-3

	Equipment 1	Equipment 2	Equipment 3
Job A	0	0	0
Job B	5	15	10
Job C	15	25	25

Table 10-4

	Equipment 1	Equipment 2	Equipment 3
Job A	0	0	0
Job B	0	10	5
Job C	0	10	10

Analysis by Jobs

The best choice (referring to Table 10-4) is to pick:

Truck 1 or 2 or 3 for job A
Truck 1   for job B
Truck 1   for job C

Job A can use any of the three trucks without loss, but job B and job C will choose optimally only if each uses truck 1; since this is impossible, either B or C must use a truck other than 1. Since you will lose less if you use another truck for job B (truck 3 at a loss of $5) than if you use another truck for job C (truck 2 or 3 at a loss of $10), you make your choices as follows:

Truck 1 for job C, truck 3 for job B, truck 2 for job A

Cost = $70 + $80 + $60 = $210

Analysis by Machines

The best choice is to pick

Job A, B, or C for truck 1
Job A   for truck 2
Job A   for truck 3

Since either truck 2 or truck 3 must take another job, you will lose the least if you give truck 3 job B (loss = $5) rather than giving truck 2 job B or C (loss = $10) or:

Job C to truck 1, job B to truck 3, job A to truck 2

Cost = $70 + $80 + $60 = $210 (as above)

What Did You Do?

You see intuitively what you did, but you must know *systematically* what you did, so that you can solve a big problem step by step without constantly trying to puzzle out where you are.

Analysis by Jobs said "The *least* loss I can take—by selecting A,2 or A,3—is $5. (I can lose more than $5 if I select between 2 and 3

PROGRAMMING

unwisely, but *at this step* in the proceedings, I am assured of a loss of *at least* $5 if I do not select A,1.)"

In Analysis by Machines, above, you said "The least loss I can take—by selecting 1,B or 1,C—is $5. (If I avoid selecting 1,A *at this point* I am sure I will take a loss of *at least* $5—perhaps more, but no less.)"

It would be most foolish to put job A in truck 1, for then *both* other jobs would have to make do with higher-cost trucks. In this case, you would incur a loss of $5 as before (job B with truck 3), *plus* a loss of $10 (job C with truck 2), or $15. Even if cell C,2 were no higher than cell B,3 (your selection technique on pages 175 and 176 ensures that it can't ever be lower) your loss would have to be *twice* $5, at least.

You can make an automatic rule at this point, referring to Table 10-4:
1. Draw a line through all the zeros in the top row.*
2. Draw a line through all the zeros in the left column.*
3. Circle the smallest value in any cell that does not contain either of these lines.
4. If you select a job-truck combination represented by any cell through which a line is drawn, your *least* loss will be the amount in that cell plus the circled value (and it might be more).
5. If you select a job-truck combination represented by any cell in which two lines intersect, your *least* loss will be the amount in that cell plus *twice* the circled value (and it might be more).
6. Each of the other cells already contains the loss that will be incurred from using its job-truck combination, and therefore each remains unchanged.

The First Solution in Summary

Accomplishing these six steps changes the loss table on the left to the one on the right in Table 10-5:

The Second Solution

The first solution does not give the true opportunities for loss in selecting one combination over another, because it treats equipment choice and

* The general method requires that you draw the *least number of lines* that will go through all zero cells.

Table 10-5

	1	2	3			1	2	3
A	0	0	0		A	10	5	5
B	0	10	⑤	becomes	B	5	10	5
C	0	10	10		C	5	10	10

Table 10-6 Recapitulates all your steps to date

$$\begin{Bmatrix} \text{Initial cost} \\ \text{matrix} \end{Bmatrix} \text{to} \begin{Bmatrix} \text{Loss table} \\ \text{for equip-} \\ \text{ment-choice} \end{Bmatrix} \text{to} \begin{Bmatrix} \text{Combined} \\ \text{equipment-} \\ \text{job-choice} \\ \text{loss table} \end{Bmatrix} \text{to} \begin{Bmatrix} \text{Adjusted} \\ \text{loss table} \\ \text{initial} \\ \text{solution} \end{Bmatrix}$$

$$\begin{pmatrix} \$55 & 60 & 70 \\ 60 & 75 & 80 \\ 70 & 85 & 95 \end{pmatrix} \text{to} \begin{pmatrix} 0 & 5 & 15 \\ 0 & 15 & 20 \\ 0 & 15 & 25 \end{pmatrix} \text{to} \begin{pmatrix} 0 & 0 & 0 \\ 0 & 10 & 5 \\ 0 & 10 & 10 \end{pmatrix} \text{to} \begin{pmatrix} 10 & 5 & 5 \\ 5 & 10 & 5 \\ 5 & 10 & 10 \end{pmatrix}$$

job choice independently. Repeating the solution several times is necessary, each time using this approximation method to get closer to the answer, until an optimum solution is reached.

The second solution (Table 10-7) is a repeat of the first, except that it starts with the "adjusted loss table" from the initial solution:

You don't go through all the steps of Table 10-6, because when you get to the second step you have a solution: there are three zeros not in the same row or same column (A,2 and B,3 and C,1). If you had not reached a solution at this point, you would have continued your second solution, then a third, and so on until a solution appeared.

A Short-cut Algorithm

In Table 10-6, when you went from the third to the fourth step, you:
1. Added twice the circled value to the double-lined cell(s)
2. Added the circled value to the single-lined cells
3. Added nothing to the nonlined cells

Then in Table 10-7, when you went from the first to the second step, you subtracted the circled value from *every* cell.

You can save a step by combining these two operations:

Short cut:
1. Add the circled value to the double-lined cell(s).
2. Add nothing to the single-lined cells.
3. Subtract the circled value from the nonlined cells.

Table 10-7

$$\begin{Bmatrix} \text{New initial} \\ \text{matrix} \\ \text{(adjusted loss} \\ \text{table from initial} \\ \text{solution)} \end{Bmatrix} \text{to} \begin{Bmatrix} \text{Loss table} \\ \text{for equipment} \\ \text{choice} \end{Bmatrix}$$

$$\begin{pmatrix} 10 & 5 & 5 \\ 5 & 10 & 5 \\ 5 & 10 & 10 \end{pmatrix} \text{to} \begin{pmatrix} 5 & 0 & 0 \\ 0 & 5 & 0 \\ 0 & 5 & 5 \end{pmatrix}$$

Summary of the assignment method

Here, summarizing in brief what has been described above, are the systematic rules for accomplishing one solution:
1. From the initial problem statement (or from the result of the previous solution), make an equipment-choice loss table.
2. From the result of step 1, make a job-choice loss table (actually a job/equipment-choice loss table).
3. On the table resulting from step 2, draw the least number of lines that will mark out all zero-value cells.
4. Circle the smallest value in any nonlined cell.

Add this circled number to the value in each cell located at an intersection of lines. *Subtract* this circled number from the value in each unlined cell.

If you do not have a solution, repeat the process as many times as necessary to obtain a solution.

A sample problem

The Problem*

You have a shop with six machines, and you have six jobs which you wish to assign to machines in the optimal way. Each machine can do any of the jobs, at varying costs as shown in Table 10-8.

The Solution

Solution of this problem by the assignment method is as shown in Table 10-9.

* A limitation of the assignment method is that you must have the same number of machines as jobs.

Table 10-8

Job no.	Machine no.					
	1	2	3	4	5	6
A	$250	260	285	290	310	340
B	220	200	230	260	270	300
C	220	260	280	240	220	290
D	230	290	300	280	260	320
E	350	320	340	395	380	400
F	250	230	265	280	290	315

Table 10-9

First solution

Machine-choice loss table

	1	2	3	4	5	6
A	0	10	35	40	60	90
B	20	0	30	60	70	100
C	0	40	60	20	0	70
D	0	60	70	50	30	90
E	30	0	20	75	60	80
F	20	0	35	50	60	85

Job/machine-choice loss table

	1	2	3	4	5	6
A	0	10	15	20	60	20
B	20	0	10	40	70	30
C	0	40	40	0	0	0
D	0	60	50	30	30	20
E	30	0	0	55	60	10
F	20	0	15	30	60	15

Second solution

Machine-choice loss table

	1	2	3	4	5	6
A	0	10	5	10	50	10
B	20	0	0	30	60	20
C	10	50	40	0	0	0
D	0	60	40	20	20	10
E	40	10	0	55	60	10
F	20	0	5	20	50	5

(Job/machine-choice loss table will be unchanged from machine-choice loss table, since each column contains at least one zero.)

Third solution

	1	2	3	4	5	6
A	0	10	5	5	45	5
B	20	0	0	25	55	15
C	15	55	45	0	0	0
D	0	60	40	15	15	5
E	40	10	0	50	55	5
F	20	0	5	15	45	0

(same)

Fourth (final) solution

	1	2	3	4	5	6
A	0	10	5	0	40	5
B	20	0	0	20	50	15
C	20	60	50	0	0	5
D	0	60	40	10	10	5
E	40	10	0	45	50	5
F	20	0	5	10	40	0

Solution:

A,4	$290	Optimum assignment and cost
B,2	200	
C,5	220	
D,1	230	
E,3	340	
F,6	315	
	$1,595	

THE TRANSPORTATION PROBLEM

Application

You have seen that the assignment method is applicable to a certain form of problem: you have several tasks and several equipments, each

PROGRAMMING

equipment can perform each task, and the costs are different. The transportation method applies to another specialized form of problem: you have several origins and destinations for shipment of material, there are limits on the amount of material available at each origin and required at each destination, and shipping costs for each route are different.

Suppose you have the mundane problem of collecting garbage efficiently. Your city has four garbage routes, each with an assembly area, and three incinerators; the amounts generated daily at each assembly area, and the maximum disposal capacity of each incinerator are shown in Table 10-10. Transportation costs between each assembly area and each incinerator differ, as do disposal costs at each incinerator; these summed costs per truckload are stated in the cells of Table 10-10. You must determine the optimum routing.*

Initial solution

The essence of programming is to start with a workable solution without regard for economy and proceed through successive improvements to an optimum solution. The only advantage of starting with a reasonably economical solution is saving of computation time; for no matter how uneconomical your first solution, eventually you will reach the optimum. If intuition suggests a reasonable initial allocation, use it—but intuition is not very helpful in a transportation problem with many sources and destinations. If your goal is improvement of an ongoing operation, start with the existing allocation.

If you don't have an existing solution and don't trust intuition, the

---

* The problem would be solved in the same way if, instead of generating garbage, the four areas were generating demand for a product; and instead of burning garbage, the three incinerators were three plants manufacturing the product at various production and shipping costs.

Table 10-10

Transportation cost matrix, dollars per truckload

Incinerators		Assembly areas				Slack
No.	Capacity	1	2	3	4	
A	100	$75	$56	$68	$62	0
B	60	52	70	60	60	0
C	80	58	62	54	58	0
	240	40	50	60	70	20

Total disposal capacity | Total daily garbage production to be picked up and burned | Excess disposal capacity

Table 10-11

Cell		1	2	3	4	Slack
A	100	$75 40	$56 50	$68 10	$62	$0
B	60	$52	$76	$60 50	$60 10	$0
C	80	$58	$62	$54	$58 60	$0 20
	240	40	50	60	70	20

"northwest corner" rule will provide an initial solution: load the upper left cell (A,1) with all it will take, go right and fill up each cell until you reach capacity, and continue this stair-step operation until you've assigned all inputs to an output. (It is as if you were to lift the southeast corner of the pinball machine and let the balls fill up all the slots in the opposite corner.)

A northwest corner initial solution is displayed in Table 10-11. To get it, you:

> Put 40 loads in cell A,1 (up to column limit at bottom).
> Put 50 loads in cell A,2 (up to column limit at bottom).
> Put 10 loads in cell A,3 (brings row total to its limit of 100).
> Put 50 loads in cell B,3 (brings column total to its limit of 60).
> Put 10 loads in cell B,4 (brings row total to its limit of 60).
> Put 60 loads in cell C,4 (brings column total to its limit of 70).
> Put the remainder in the slack cell of row C.

Note that this solution leaves incinerator C underutilized by 20 loads per day.

A better solution

Concept of improving the solution

Of 15 cells you might have used (counting the 3 slack cells which indicate less than capacity utilization), your initial solution uses only 7. You must consider, for each unused cell in turn, whether you could reduce expenses by using it.

Consider cell B,1. If you add *one* truckload there, your costs will change +$52. To keep the column 1 total at 40 (the daily production of route 1), you must take one truckload from cell A,1, changing your costs further by −$75. To keep the row A total at 100 (the daily capacity for incinerator A), you must add 1 truckload to A,2 or A,3. (If you select the former and raise its cell total to 51, there is no other cell in that column that can give up 1 to bring the total back down to 50.

Table 10-12

Cell	1	2	3	4	5
A	75 / 40	56 / 50	68 / 10	62 / (-6)	0 / (-10)
B	52 / (-15)	70 / (+22)	60 / 50	60 / 10	0 / (-2)
C	58 / (-7)	62 / (+16)	54 / (-4)	58 / 60	0 / 20

Daily cost $13,560

So it has to be A,3.) Adding 1 truckload to A,3 changes your costs further by +$68. To keep the column 3 total at 60, you must take 1 truckload from B,3, changing your costs further by −$60. The overall cost change if you add 1 truckload to B,1 is:

$$\$52 - \$75 + \$68 - \$60 = -\$15$$

You can reduce your overall costs $15 for every truckload you add to cell B,1, to a total of 40 loads. At this point, cell A,1 will be empty, and you can add no more.

Before you decide to make this change, you should analyze all the unused cells to see if you can do even better elsewhere. The path for cell C,1 would go as follows:

$$+(C,1) - (A,1) + (A,3) - (B,3) + (B,4) - (C,4)$$
$$+\$58\ -\ \$75\ +\ \$68\ -\ \$60\ +\ \$60\ -\ \$58\ =\ -\$7$$

Table 10-12 shows, in parentheses, the results of calculating unit gains or losses for each unused cell. (Since you are considering the addition of only *one cell at a time*, on each iteration your path for calculating the cost of adding one empty cell cannot involve another empty cell.)

Second Solution

B,1 is the most advantageous cell to bring into the allocation. You therefore bring it in at the maximum possible amount of 40 truckloads, make the resultant changes in other cells, and you have a second solution, Table 10-13, on which the circled allocations are those you changed in your second solution. This change saves you 40 × $15 = $600.

Table 10-13

Cell	1	2	3	4	Slack
A	75 / ⓪ (+5)	56 / 50	68 / ㊿	62 / (-6)	0 / (-10)
B	52 / ㊵	70 / (+22)	60 / ⑩	60 / 10	0 / (-2)
C	58 / (+8)	62 / (+16)	54 / (-4)	58 / 60	0 / 20

Daily cost $12,960

Table 10-14

Cell	1	2	3	4	Slack	
A	75 (+15)	56 / 50	68 / ⓐ40	62 (+4)	0 / ⑩	Daily cost $12,860
B	52 / 40	70 (+22)	60 / ⓐ20	60 / 0 (+10)	0 (+8)	
C	58 (-2)	62 (+6)	54 (-14)	58 / ⓐ70	0 / ⑩	

Third Solution

The best addition is cell A,5 (−$10) which can follow path +(A,5) − (A,3) + (B,3) − (B,4) + (C,4) − (C,5) and add 10 truckloads at a saving of 10 × $10 = $100. See Table 10-14.

Fourth Solution

The best addition is cell C,3 which can add 10 truckloads at a saving of 10 × $14 = $140. See Table 10-15.

Fifth Solution

The best addition is cell A,4, which can add 30 truckloads at a saving of 30 × $10 = $300. See Table 10-16.

Table 10-15

Cell	1	2	3	4	Slack	
A	75 (+15)	56 / 50	68 / ㉚	62 (-10)	0 / ⑳	Daily cost $12,720
B	52 / 40	70 (+22)	60 / 20	60 (-4)	0 (+8)	
C	58 (-12)	62 (+20)	54 / ⑩	58 / 70	0 / ⓪ (+14)	

Table 10-16

Cell	1	2	3	4	Slack	
A	75 (+25)	56 / 50	68 / ⓪ (+10)	62 / ㉚	0 / 20	Daily cost $12,420
B	52 / 40	70 (+12)	60 / 20	60 (-4)	0 (-2)	
C	58 (+12)	62 (+10)	54 / ㊵	58 / ㊵	0 (+4)	

PROGRAMMING

Table 10-17

Cell	1	2	3	4	Slack
A	75 / (+21)	56 / 50	68 / (+10)	62 / 30	0 / 20
B	52 / 40	70 / (+16)	60 / (+4)	60 / 20	0 / (+2)
C	58 / (+8)	62 / (+10)	54 / 60	58 / 20	0 / (+4)

Daily cost $12,340

Sixth Solution

The best addition is cell B,4, which can add 20 truckloads at a saving of 20 × $4 = $80. See Table 10-17.

At this point all unit price changes are plus. No further improvement is possible, and you have reached an optimum solution.

LINEAR PROGRAMMING

The previous methods were applicable only to special types of resource allocation problems. A general method which does not require such special conditions is the linear programming method.

Determinate linear systems

Two-product, Two-machine Problem

Suppose you are manager of a factory which makes two types of metal boxes and has two machines—a punch press and a spot welder. Each machine costs the same amount to operate, and the manufacture of either box requires time on both machines as follows:

No. 1 boxes: 3 minutes on welder and 6 minutes on press
No. 2 boxes: 3 minutes on welder and 2 minutes on press

If x_1 and x_2 represent the number of No. 1 and No. 2 boxes manufactured per hour, respectively, then you will keep the welder busy for 60 minutes by any combination of x_1 and x_2 satisfying the following equation:

$$3x_1 + 3x_2 = 60 \tag{10-1}$$

If you manufacture 9 of the No. 1 boxes ($x_1 = 9$) at 3 minutes each, you use 27 minutes; in the 33 minutes remaining for No. 2 boxes at 3 minutes each you can make 11 ($x_2 = 11$). A similar equation for the press would be:

$$6x_1 + 2x_2 = 60 \tag{10-2}$$

Your problem is to determine how many boxes of each type you should make to keep both machines working to capacity, which means finding values for x_1 and x_2 that satisfy both equations simultaneously.

When a high school algebra student looks at such a pair of equations, he hunts for a simple path to a solution in the way a wrestler seeks an opening. He notices that Eq. (10-1) can be divided through by 3, making it $x_1 + x_2 = 20$, which he rearranges to set $x_2 = 20 - x_1$. He substitutes that in Eq. (10-2) to make it $6x_1 + 40 - 2x_1 = 60$, solves that to get $x_1 = 5$, which he substitutes back in Eq. (10-1) to get $x_2 = 15$. In other words, make 5 of the No. 1 boxes and 15 of the No. 2 every hour.

Quick, isn't it? But a method hand-tailored for a particular problem won't do. You need a systematic method, no matter how seemingly clumsy, that works for any problem—and that the computer can understand.

Solution by Algebraic Elimination

The following method of solving this simple problem will strike you as idiotic, for it will seem to take forever. The systematic approach works when you can't see the trail, and thus it can handle massive problems. The solution is carried forward in two columns—the left uses actual numbers from the example, and the right uses generalized symbols as a necessary prelude to building a general method or algorithm. (It isn't necessary to study these operations in detail, but you ought to see generally what is being done.)

1. *The equations.*

$$3x_1 + 3x_2 = 60 \quad\text{or}\quad A_{11}x_1 + A_{12}x_2 = b_1 \quad (10\text{-}3)$$
$$6x_1 + 2x_2 = 60 \qquad\qquad A_{21}x_1 + A_{22}x_2 = b_2 \quad (10\text{-}4)$$

2. *Elimination of x_1.*

   a. Divide Eq. (10-3) by 3
   (use 3 as "pivot"):

$$\frac{1}{3} \times 3x_1 + \frac{1}{3} \times 3x_2 = \frac{1}{3} \times 60 \qquad (10\text{-}5)$$

   Divide Eq. (10-3) by A_{11}
   (use A_{11} as "pivot element"):

$$\frac{A_{11}}{A_{11}} x_1 + \frac{A_{12}}{A_{11}} x_2 = \frac{b_1}{A_{11}} \qquad (10\text{-}5)$$

   b. *Multiply Eq. (10-5) by 6:*   *Multiply Eq. (10-5) by A_{21}:*
   (so that the coefficient of x_1 in the two equations will be equal,

PROGRAMMING

and you can subtract)

$$\frac{6}{3} \times 3x_1 + \frac{6}{3} \times 3x_2 = \frac{6}{3} \times 60 \qquad \frac{A_{21} \times A_{11}}{A_{11}} x_1$$

$$+ \frac{A_{21} \times A_{12}}{A_{11}} x_2 = \frac{A_{21} \times b_1}{A_{11}}$$

c. *Then subtract b from Eq. (10-4):*

$$6x_1 + 2x_2 = 60 \qquad\qquad A_{21}x_1 + A_{22}x_2 = b_2$$
$$\text{minus} \qquad\qquad\qquad \text{minus}$$
$$\frac{6}{3} \times 3x_1 + \frac{6}{3} \times 3x_2 \qquad \frac{A_{21}A_{11}}{A_{11}} x_1 \frac{A_{21}A_{12}}{A_{11}} x_2 = \frac{A_{21}b_1}{A_{11}}$$
$$= \frac{6}{3} \times 60 \qquad\qquad \text{equals}$$
$$\text{equals} \qquad\qquad 0x_1 + \left(A_{22} - \frac{A_{21}A_{12}}{A_{11}}\right) x_2$$
$$0x_1 + [2 - \tfrac{6}{3}(3)]\, x_2$$
$$= 60 - \tfrac{6}{3}(60) \qquad\qquad = b_2 - \frac{A_{21}}{A_{11}} b_1 \qquad (10\text{-}6)$$

3. *Solving for* x_2.

  a. *Divide Eq. (10-6) by* $[2 - \tfrac{6}{3}(3)]$:   *Divide Eq. (10-6) by*
  $(-\tfrac{1}{4})(-4x_2) = (-\tfrac{1}{4})(-60)$
  $$\left(A_{22} - \frac{A_{21}}{A_{11}} A_{12}\right):$$
  $$\frac{A_{22} - \dfrac{A_{21}}{A_{11}}(A_{12})x_2}{A_{22} - \dfrac{A_{21}}{A_{11}} A_{12}}$$
  $$= \frac{\left(b_2 - \dfrac{A_{21}}{A_{11}} b_1\right)}{A_{22} - \dfrac{A_{21}}{A_{11}} A_{12}}$$

  b. *And simplify:*

  $$0x_1 + x_2 = 15 \qquad 0x_1 + x_2 = \frac{A_{11}b_2 - A_{21}b_1}{A_{11}A_{22} - A_{21}A_{12}} \qquad (10\text{-}7)$$
  or $\quad x_2 = 15 \quad$ or $\quad x_2 = \dfrac{A_{11}b_2 - A_{21}b_1}{A_{11}b_{22} - A_{21}A_{12}}$

4. *Solving for* x_1. You want to multiply Eq. (10-7) by coefficient of x_2 in Eq. (10-5). When you subtract it from Eq. (10-5), x_2 will disappear and the coefficient of x_1 will be $1 - 0 = 1$, and you can solve for x_1 directly.

a. *Multiply Eq. (10-7) by $\frac{1}{3}(3)$:* *Multiply Eq. (10-7) by A_{12}/A_{11}:*

$$0x_1 + \frac{1}{3} \times 3x_2 = \frac{1}{3} \times 3 \times 15 \qquad 0x_1 + \left(\frac{A_{12}}{A_{11}}\right)x_2$$
$$= \frac{A_{12}}{A_{11}}\left(\frac{A_{11}b_2 - A_{21}b_1}{A_{11}A_{22} - A_{21}A_{12}}\right)$$

b. *Then subtract a from Eq. (10-4):*

$$x_1 + x_2 = 20 \qquad x_1 + \frac{A_{12}}{A_{11}}x_2 = \frac{b_1}{A_{11}}$$

minus minus

$$0x_1 + \frac{1}{3} \times 3x_2 \qquad 0x_1 + \frac{A_{12}}{A_{11}}x_2 = \frac{A_{12}}{A_{11}}\left(\frac{A_{11}b_2 - A_{21}b_1}{A_{11}A_{22} - A_{12}A_{21}}\right)$$
$$= \frac{1}{3} \times 3 \times 15$$

equals equals

$$x_1 + 0x_2 = 5 \qquad x_1 + 0x_2 = \frac{b_1}{A_{11}}$$
$$- \frac{A_{12}}{A_{11}}\left(\frac{A_{11}b_2 - A_{21}b_1}{A_{11}A_{22} - A_{12}A_{21}}\right) \quad (10\text{-}8)$$

$$x_1 = 5 \qquad x_1 = \frac{b_1}{A_{11}} - \frac{A_{12}}{A_{11}}\left(\frac{A_{11}b_2 - A_{21}b_1}{A_{11}A_{22} - A_{12}A_{21}}\right)$$

Solution by Algorithm Table

The right side of the above sequence can be put into abbreviated tabular form, which describes exactly the same operations but omits intermediate steps. The algorithm table, Table 10-18, uses algebraic notation for reciprocals; for example: $(A_{11})^{-1}$ means the same as $1/A_{11}$, so that multiplying by $(A_{11})^{-1}$ is the same as dividing by A_{11}.

Table 10-18

Eq^n	Operation	Coefficient X_1	Coefficient X_2	b
(I)		A_{11}	A_{12}	b_1
(II)		A_{21}	A_{22}	b_2
(III)	(I) $\div A_{11}$ or (I)$(A_{11})^{-1}$	1	$\dfrac{A_{12}}{A_{11}}$	b_1/A_{11}
(IV)	(II) $-$ (III)(A_{21})	0	$(A_{22} - \dfrac{A_{21}}{A_{11}}(A_{12}))$	$b_2 - \dfrac{A_{21}}{A_{11}}b_1$
(V)	(IV) $\times \left[A_{22} - \dfrac{A_{21}}{A_{11}}(A_{12})\right]^{-1}$	0	1	$\dfrac{A_{11}b_2 - A_{21}b_1}{A_{11}A_{22} - A_{21}A_{12}} = X_2$
(VI)	(III) $-$ (V) $\times \dfrac{A_{12}}{A_{11}}$	1	0	$\dfrac{b_1}{A_{11}} - \dfrac{A_{12}}{A_{11}}\left[\quad\diagup\quad\right]$ $\dfrac{A_{11}b_2 - A_{21}b_1}{A_{11}A_{22} - A_{21}A_{12}} = X_1$

Table 10-19

Eq^n	Operation	x_1	x_2	b	Σ
(I)		A_{11}	A_{12}	b_1	S_1
(II)		A_{21}	A_{22}	b_2	S_2
(III)	$(I) \times (A_{11})^{-1}$	1	$A_{12.1}$	$b_{1.1}$	$S_{1.1}$
(IV)	$(II) - (III) \times (A_{21})$	0	$A_{22.1}$	$b_{2.1}$	$S_{2.1}$
(VI)	$(III) - (V) \times (A_{12.1})$	1	0	$b_{1.2}$	$S_{1.2}$
(V)	$(IV) \times (A_{22.1})^{-1}$	0	1	$b_{2.2}$	$S_{2.2}$

Now rephrase this table somewhat, as shown in Table 10-19.
1. A coefficient used as first pivot is A_{11} (or A_{21}, etc.) Call the coefficient used as second pivot (after one iteration) $A_{21.1}$, $A_{12.1}$, etc. Renumber entries in the above table accordingly.
2. Now also interchange entries (V) and (VI); the equations will continue to be in pairs, after each iteration, with x_1 first and x_2 second.
3. Add a "sum" column, as an arithmetic check (to be explained later).

Now work your problem using Table 10-20.

Before you leave the two-product, two-machine problem, look at it graphically in Fig. 10-1. The horizontal axis represents the number of No. 1 boxes made per hour, and the vertical axis the number of No. 2 boxes. When $x_2 = 0$ (no No. 2 boxes are being made), the welder can turn out 20 of the No. 1 boxes ($x_1 = 20$); and when $x_1 = 0$, $x_2 = 20$. A line connecting these two points contains all combinations of No. 1 and No. 2 boxes that will "satisfy" the welder equation (keep it busy 60 minutes an hour). The other line is drawn the same way, and represents

Table 10-20

	Eq	Operation	x_1	x_2	b	Σ^*
1st	(I)		3	3	60	66
2nd	(II)		6	2	60	68
3rd	(III)	$(I) \times (1/3)$	1	1	20	22
4th	(IV)	$(II) - (III) \times (6)$	$6-1 \times 6 = 0$	$2-1 \times 6 = -4$	$60 - 20 \times 6 = -60$	$68 - 22 \times 6 = -64$
6th	(VI)	$(III) - (V) \times (1)$	$1-0=1$	$1-1=0$	$20-15=5$	$22-16=6$
5th	(V)	$(IV) \times (-1/4)$	0	$(-4) \div (-4) = 1$	$-60/-4 = 15$	$-64/-4 = 16$

*Elements in the "Σ" column are calculated like elements in other columns, then checked by adding across the row

This x_1 is x_2

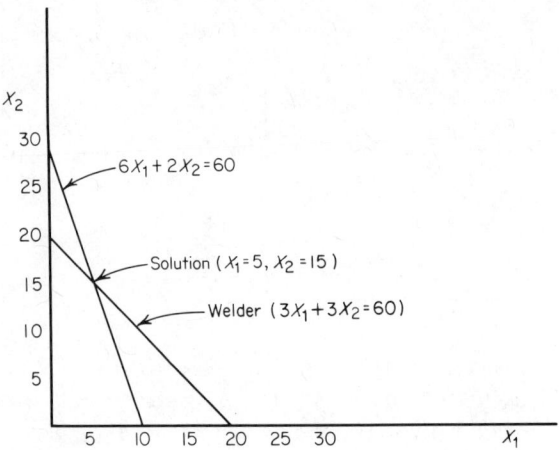

Fig. 10-1

all possible combinations of x_1 and x_2 that will keep the press busy. Their intersection therefore has to be the mix of boxes—and the only mix—that will keep both machines busy.

This was a very special problem because it had a single solution that would keep all your machines busy. Usually you aren't so fortunate.

Indeterminate linear systems

Two-product, Three-machine Problem

You are running the same factory, but competition forces you to improve the finish of your product. You add a third machine, a plater, which requires 8 minutes for each No. 1 box. It isn't used for No. 2 box, which you continue to turn out for the plebeian trade. This gives you three equations instead of two:

Welder:	$3x_1 + 3x_2 = 60$	(10-9)
Press:	$6x_1 + 2x_2 = 60$	(10-10)
Plater:	$8x_1 + 0x_2 = 60$	(10-11)

Graphical Representation

You can represent these three machines, and the product mix each can handle in 60 minutes, by adding a line for the plater equation to the previous graph, as shown in Fig. 10-2. Since the equation $8x_1 = 60$ is satisfied by the value $x_1 = 7\frac{1}{2}$ for any value of x_2, it is plotted as a vertical line crossing the horizontal axis at $7\frac{1}{2}$.

A problem arises immediately, as you inspect the graph. If you keep the welder and the press busy, as before ($x_1 = 5$, $x_2 = 15$), the plater is underutilized; it can handle 7½ of the No. 1 boxes, and you are making only 5. If you keep the press and the plater busy, you underutilize the welder. And you *can't* keep the welder and plater busy, because their intersection lies outside the press equation line—you would be scheduling the press for more than 60 minutes an hour, or more output than it actually has. You can't have any product mix that lies outside the five-sided figure shaded on the graph—and that means you won't find any mix that will keep all three machines busy.

The three equations are not *requirements* that you must satisfy, then, but *limits* that you cannot exceed. You can't schedule a product mix that will exceed 60 minutes on any machine, but it is perfectly feasible to have a product mix that requires less than 60 minutes. The equations, rewritten to express this requirement, become:

$$3x_1 + 3x_2 \leq 60 \qquad (10\text{-}12)$$
$$6x_1 + 2x_2 \leq 60 \qquad (10\text{-}13)$$
$$8x_1 + 0x_2 \leq 60 \qquad (10\text{-}14)$$

(The symbol $\leq$ means "less than or equal to.")

These are not equations, but inequalities, and they cannot be solved algebraically. They can be made into equations, however, by putting into each equation a variable to represent the minutes of unused time, or slack time, that would be required to bring the total up to 60 minutes again. If only 5 No. 1 boxes were made, requiring only 40 minutes per hour on the plater, we could make Eq. (10-14) an equation again by saying $8x_1 + S = 60$, and when x_1 equals 5, S will equal 20 minutes of slack time. One such variable is needed in each equation, to provide for whatever number of minutes of slack time that machine has for any

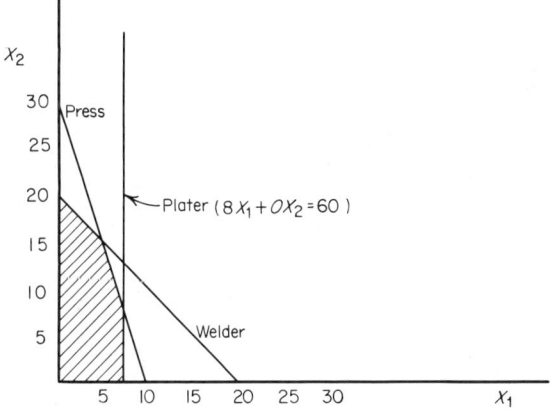

Fig. 10-2

product mix. Since variables x_1 and x_2 have been used as basic variables, use S_3, S_4, and S_5 as the three slack variables.

The inequalities become proper equations, written as follows:

$$3x_1 + 3x_2 + S_3 = 60 \tag{10-15}$$
$$6x_1 + 2x_2 + S_4 = 60 \tag{10-16}$$
$$8x_1 + 0x_2 + S_5 = 60 \tag{10-17}$$

Now you have three equations and five unknowns; and in high school algebra you couldn't solve a system of simultaneous equations if there were more unknowns than equations. Actually there are infinitely many solutions—anywhere within the shaded area or "feasibility envelope." (One feasible solution is at the origin, where x_1 and x_2 are both zero; you produce this product mix every Sunday.) But which of this infinite number of solutions is optimum?

The Objective Function

Without more information, you have no basis for choice. You need to know the objective of the game—the "objective function." Since the goal of your factory is profit, you must specify what profit a unit of each product brings. Your bookkeeper tells you that each No. 1 box brings

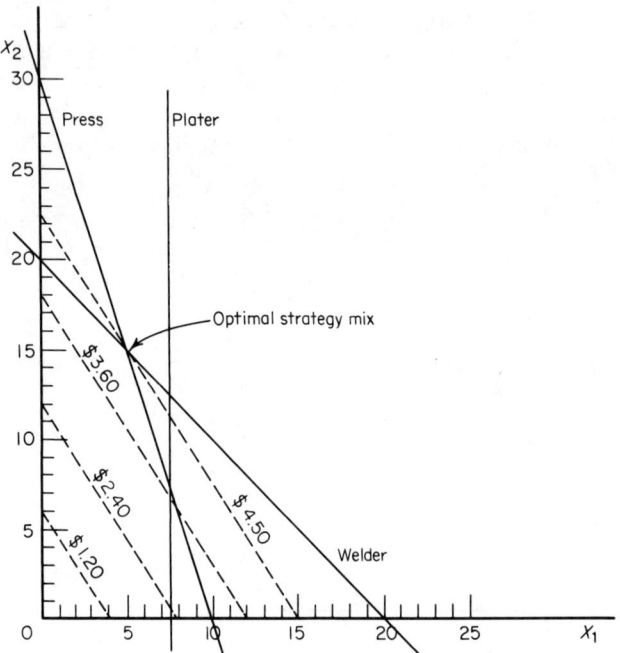

Fig. 10-3

30 cents net profit and each No. 2 box brings 20 cents net profit.* Since 4 of the No. 1's or 6 of the No. 2's bring the same $1.20 profit, you can draw a line on the previous graph connecting these two quantities as in Fig. 10-3—and any mix of No. 1 and No. 2 boxes lying along that line will bring $1.20 profit. As you continue to increase output in this way by drawing successive lines parallel to this line but farther from the original, you see that the feasibility envelope begins to limit your possible product mix.

With the line connecting 15 on the x_1 axis and $22\frac{1}{2}$ on the x_2 axis you reach the maximum profit of $4.50—and the only way you can attain it is to schedule your hourly production precisely where your "equal-profit line" reaches the point in the feasibility envelope farthest from the origin: where $x_1 = 5$ and $x_2 = 15$.

Your optimal strategy is to make 5 of the No. 1 boxes and 15 of the No. 2 each hour.

Your hourly profit is the value obtained by inserting these values into your "profit function" or "objective function" equation:

$$\text{Hourly profit} = 30¢ \times x_1 + 20¢ \times x_2$$
$$= 30¢ \times 5 + 20¢ \times 15 = 450¢ \text{ or } \$4.50$$

The simplex method

The Need for a Systematic Technique

If the above example were typical, there would be no need for further techniques: you could solve any problem graphically. But real problems are far too complex for this. Imagine adding a third box, requiring an additional x_3 axis perpendicular to the paper; the capacity limitations and the objective function now would be planes—like so many triangular pieces of plywood enclosing a many-sided space in the corner of a room—and you would need all your spatial perception to find the optimum vertex. Make that four boxes or more and you couldn't draw a graph at all. You need a method which follows the concept of the simple graphical solution, but works on the elaborate situations of real life.

Concept of the Simplex Method

A manager says, "Maximize profit, subject to staying within capacity limits of all departments." A linear programmer says, "Maximize† this

---

* You will argue that profit is affected by volume, so it doesn't remain at 30 cents for any number produced, and you are quite right of course; but this is a limitation of linear programming—it is linear. However, unit profits tend to be linear over usual ranges of efficient operations.

† He might say "minimize," if he were working with costs, say, instead of profits. In this case he would simply change all the signs in his objective function, and proceed in the usual way.

objective function, subject to these constraints" and adds slack variables to turn the constraints into equations.

These expressions, after adding the slack variables, would look like this for our two-box problem:

Constraints:
$$3x_1 + 3x_2 + S_3 = 60 \quad (10\text{-}18)$$
$$6x_1 + 2x_2 + S_4 = 60 \quad (10\text{-}19)$$
$$8x_1 + 0x_2 + S_5 = 60 \quad (10\text{-}20)$$

Objective function: $\quad 30x_1 + 20x_2 + 0S_3 + 0S_4 + 0S_5 = F$

Keep in mind what the variables mean. Equation (10-18) states that, for the welder:

| The number of No. 1 boxes made per hour times 3 | + | the number of No. 2 boxes made per hour times 3 | + | the number of minutes of idle time per hour | = | 60 minutes each hour |

Equations (10-19) and (10-20) say the same thing for the press and the plater, with S_4 the idle minutes for the press and S_5 the idle minutes for the plater.

All five variables go into the objective function; since slack time makes no profit, the coefficients of the three slack variables must be 0.

You can see from considering the shaded area of Fig. 10-2 that the optimum point has to be at some corner (or vertex) for your two-dimensional problem.* The branch of mathematics known as convex sets can show that this is true for all cases, no matter how many "dimensions." And you can prove to yourself that each of these corners represents a situation where *two of the five variables are equal to zero*, leaving three variables for the three equations—which the algebraist can solve.

Set $x_1 = 0$ and $x_2 = 0$, for instance. Physically you can see that this puts you at the lower left vertex, with zero production of any boxes. Mathematically, this is as though you had erased x_1 and x_2 from three equations, leaving $S_3 = 60$, $S_4 = 60$, and $S_5 = 60$; physically, you have 60 minutes of slack time per hour on each of the three machines. You see physically that you make zero profit; mathematically, when you put the above values for your five variables into the objective function equation, the result is zero.

Move counterclockwise to the lower right vertex. You have brought x_1 into the action; it isn't zero any more, but has a value of $7\frac{1}{2}$.

---

* Had your equal-profit line been parallel to one of the constraint lines—say, parallel to the welder line—then *any* mix along the top line of the shaded portion would have been optimum; but as long as the constraint lines are straight there cannot be an optimum mix that lies wholly outside of at least one vertex.

Since only three variables are allowed in the game (to go with the three equations), you must have sent some other variable back to zero—but which one? The plater now is working to capacity, so plater slack (S_5) must be zero. The programmer says you've "brought x_1 into the basis to replace S_5."

How do you know when you have increased x_1 enough to reach the corner? Physically, you can *see* that $7\frac{1}{2}$ is the limit—but in a 10-dimensional linear program you won't *see* anything. Mathematically, you take one equation at a time, set its slack variable equal to zero—and see what value of x_1 you get in each case.

In Eq. (10-18):   When S_3 goes to 0, $3x_1 = 60$, or $x_1 = \underline{20}$ (x_2 is not in the basis now, so it equals zero)
In Eq. (10-19):   When S_4 goes to 0, $6x_1 = 60$, or $x_1 = \underline{10}$
In Eq. (10-20):   When S_5 goes to 0, $8x_1 = 60$, or $x_1 = \underline{\underline{7\frac{1}{2}}}$

Which value of x_1 do you use: 20, 10, or $7\frac{1}{2}$? You must use the *lowest*, because when the first slack variable goes to zero you have reached full capacity on *that* machine and can go no further.

What you've done, mathematically, is to move from the origin along the bottom line: at $7\frac{1}{2}$ No. 1 boxes you have used up all slack time on the plater, so you can't go further (if you could, at 10 boxes you'd use up all slack time on the press; and at 20 boxes you'd use up all slack time on the welder). The plater constraint stops you, S_5 goes to 0, and x_1 comes in the basis at $7\frac{1}{2}$.

You have improved your hourly profit (to $7\frac{1}{2} \times 30\mathcal{c} = \$2.25$), but how do you know whether you have reached the optimum mix yet? In this simple example, the graph shows that you have not. In the simplex method, where you must make do with no visual aids, you have two questions at this point:
1. Have I reached the optimum yet?
2. If not, what variable do I bring into the basis next, to move me closer to the optimum?

Both of these questions are answered mathematically by looking at the objective function (in its somewhat revised form). You are out to make money: you ask whether any one of the variables now set equal to zero would increase hourly profit if it were brought into the basis at something more than zero. At this point in the operation, if you did your algebra right, your objective function would look like this:

$$F = 225 - 15/4 S_5 + 20x_2$$

Any value of x_2 will increase F in this equation, and you will therefore improve profits by bringing x_2 into the basis. (This follows because x_2

has a *positive* coefficient. In the streamlined world of linear programming you simply adopt this rule: you haven't reached an optimum as long as any nonbasic variable still has a *positive* coefficient; when all nonbasic variables in the objective function have *negative* coefficients, you are done.)

Proceeding as before, you find that the lowest value of x_2 attained from setting each of the three basic variables in turn equal to zero is $7\frac{1}{2}$, gotten when $S_4 = 0$. Physically this means that the press now is loaded to capacity (it has zero slack); graphically on Fig. 10-2 you have moved up the plater line to the point where it intersects the press line.

At this point you have worked the problem three times. The first time you solved for the profit with both x_1 and x_2 at zero, and found (not surprisingly) that there wasn't any. Next you brought one of these two variables in and repeated the solution (in programming jargon, an "iteration"). Then you did it once more—a second iteration. It would take one more iteration to reach the optimum solution: mathematically you'd bring S_5 in to replace S_3 and physically you'd put some slack in the plater schedule and run the welder to capacity, which would move you up and left along the press constraint line to the point $x_1 = 5$, $x_2 = 15$. When you reached this point, all the nonbasic variables would have negative coefficients in the objective function.

You are wondering: must I really tackle all this arithmetic? No. In the first place, the simplex method has been reduced to a systematic problem-solving technique called the "simplex tableau," which will be explained briefly in the next section. And in the second place, you won't even use that, because many computer programs can run through the tableau a million or so times faster than you can. You should know generally how the simplex tableau works (so that you'll know what you're asking the computer to do); and you should know how to set your problem up in a form the computer will accept.

The Simplex Tableau

Earlier in this chapter, an algorithm table was presented for solving determinate linear systems. Since your indeterminate problem (more unknowns than equations) reduces to repeated solving of determinate systems by setting the excess unknowns equal to zero, the simplex tableau adapts a form of this algorithm table to linear programming.

Figure 10-4 shows a simplex tableau for solving the problem graphed in Fig. 10-3. Each set of three equations constitutes an iteration—a solution for three unknowns that gives you one vertex of your feasibility envelope. Equations (I), (II), and (III) in the tableau solve the "Sunday" situation, with x_1 and x_2 set at zero to give you a starting point. Under column (8) are the answers you seek: rows (2), (3), and (4) give

Equation	Operation	Basic variables	Unit profit	$C_1 = 30$ X_1	$C_2 = 20$ X_2	$C_3 = 0$ S_3	$C_4 = 0$ S_4	$C_5 = 0$ S_5	+F b	9 (Σ)	Ratio for non/basic variables
I	Given	S_3	$C_3 = 0$	$A_{11} = 3$	$A_{12} = 3$	$A_{13} = 1$	$A_{14} = 0$	$A_{15} = 0$	$b_1 = 60$	67	$b/a*$ = 60/3 = 12.5
II		S_4	$C_4 = 0$	$A_{21} = 6$	$A_{22} = 2$	$A_{23} = 0$	$A_{24} = 1$	$A_{25} = 0$	$b_2 = 60$	69	60/6
III		S_5	$C_5 = 0$	$A_{31} = 8$	$A_{32} = 0$	$A_{33} = 0$	$A_{34} = 0$	$A_{35} = 1$	$b_3 = 60$	69	60/8**
	Change in profit			$C_{1,0} = 30*$	$C_{2,0} = 20$	$C_{3,0} = 0$	$C_{4,0} = 0$	$C_{5,0} = 0$	+$F_0 = 0$		
Pivot for first iteration: S_5 out of basis, X_1 in basis, (X_1 in because $C_{1,0}$ is largest, S_5 out because ratio is smallest so constrains X_1.)
V	$S_{j,1} = b_j - a_{j,-\text{IN}}(b_{\text{OUT}}/a_{\text{OUT}-\text{IN}})$	S_3	$C_3 = 0$	0	3	1	0	-3/8	37.5	41 1/8	37.3/3 = 12.5
VI	$S_{j,1} = b_j - a_{j,-\text{IN}}(b_{\text{OUT}}/a_{\text{OUT}-\text{IN}})$	S_4	$C_4 = 0$	0	2	0	1	-3/4	15	17 1/4	15/2 = 7.5**
IV	$S_{\text{IN},1} = b_{\text{OUT}} \div a_{\text{OUT}-\text{IN}}$	X_1	$C_1 = 30$	1	0	0	0	1/8	7.5	8 5/8	7.5/0 = 0
	Change in profit			$C_{1,1} = 0$	$C_{2,1} = 20*$	$C_{3,1} = 0$	$C_{4,1} = 0$	$C_{5,1} = -15/4$	+F_1 = +225		
Pivot for second iteration: S_4 out of basis, X_2 in basis
VIII	$S_{j,2} = b_{j,1} - a_{j,-\text{IN},1}(b_{\text{OUT},1}/a_{\text{OUT}-\text{IN},1})$	S_3	$C_3 = 0$	0	0	1	-3/2	(-3/4)	15	15 1/4	15/ 3/4 = -20**
VII	$S_{\text{IN},2} = b_{\text{OUT},1} \div a_{\text{OUT}-\text{IN},1}$	X_2	$C_2 = 20$	0	1	0	1/2	-3/8	7.5	8 5/8	7.5/-3.8
IX	$S_{j,2} = b_{j,1} - a_{j,-\text{IN},1}(b_{\text{OUT},1}/a_{\text{OUT}-\text{IN},1})$	X_1	$C_1 = 30$	1	0	0	0	1/8	7.5	8 5/8	7.5/ 1/8 = 60
	Change in profit			0	0	0	-10	$C_{5,1} = -15/4$ a_4 +F_2 = +375		
Pivot for third iteration: S_3 out of basis, S_5 in basis
X		S_5	$C_5 = 0$	0	0	4/3	-2	1	20	20 1/3	
XI		X_2	$C_2 = 20$	0	1	1/2	-1/4	0	15	16 1/4	
XII		X_1	$C_1 = 30$	1	0	-1/6	1/4	0	5	6 1/12	
	Change in profit			0	0	-10+5=-5	5-7.5=-2.5	0	+F_3 = +450 Answers		

You have finished since all coefficients in objective function are minus.

NOTE: Where "IN" and "OUT" are shown above as subscripts, they refer to the number of the variable going out of or into the basis

Fig. 10-4 Simplex tableau for two-product, three-machine problem.

you the values of S_3, S_4, and S_5—all 60 minutes slack time per hour; row (5) gives the hourly profit—$F_0 = 0$.

An explanation of the tableau's shorthand is in order. Across the top are the coefficients of the five variables in the objective function: the coefficient of x_1, called C, is 30 (you make 30 cents unit profit per No. 1 box sold), and so on. Columns (3) to (7) contain the coefficients of X_1, X_2, S_3, S_4, and S_5—for the welder, these are designated A_{11} to A_{15}, for the press, A_{21} to A_{25}, and so on. Column (8) contains the value on the right side of each constraint equation: 60 minutes available time each hour. Column (1) indicates which are the basic (nonzero) variables for this iteration; column (2) displays the coefficients in the objective function for these variables. Column (9) is an entry for checking arithmetic accuracy; you may disregard it since the computer will do your arithmetic.* Column (10) calculates which machine will use up its slack first—telling you which basic variable goes to zero, and which new variable comes into the basis.

Try the tableau to go from the starting equations to the first iteration. In column (3), row (5), the largest unit profit coefficient—30 cents—is under variable x_1; therefore it seems potentially most profitable to bring x_1 into the basis (to make some No. 1 boxes). Put an asterisk beside 30. To find which variable x_1 replaces, refer to column (10), where for each row the column (8) entry, 60, is divided by the column (2) entry (the coefficient of x_1). Double-asterisk the smallest quotient. This is the same thing you did above, when you selected $7\frac{1}{2}$ as being smaller than 10 or 20. This operation tells you to bring in the column (3) variable, x_1, and take out the row (4) entry, S_5.

Two mechanical rules need to be stated. (You could figure them by tracing through the algebra, but don't bother; the chief merit of algorithms such as this is their reduction of calculations to a simple-minded system.)

Rule for Figuring Change-in-profit Coefficients

You knew what the change-in-profit coefficients were for the initial situation, as shown in row (5)—you simply copied them from the values of c_1 to c_5 across the top. But what are they for subsequent iterations? Take $c_{2.1}$ in row (10), for example. You get it by the following formula:

$$c_{2.1} = c - \text{the sum of } (c_3 \times a_{12.1} + c_4 \times a_{22.1} + c_1 \times a_{32.1})$$
$$= 20 - \phantom{\text{the sum of }} (0 \times 3 + 0 \times 2 + 30 \times 0) = 20$$

For any iteration, in other words, the change-in-profit coefficient of a variable is

---

* The output of a poorly controlled computer can be wrong for many reasons, but arithmetic error is not one of them.

PROGRAMMING

$$\text{That variable's initial coefficient (top of tableau)} - \frac{\text{each unit profit coefficient}}{\text{for that iteration [column (2)]}} \times \frac{\text{the same-row coefficient}}{\text{of that variable}}$$

OR shorthand for the change-in-profit coefficient of, say, the jth variable in the third iteration, would be

$$c_{j.3} = c_j - \Sigma(c_{i.3} \cdot a_{ij.3})$$

Rule for Figuring Profit

You know from common sense that the profit at any vertex is 30 cents times the number of No. 1 boxes plus 20 cents times the number of No. 2 boxes. Equally true, though perhaps not so obvious, is the fact that for any iteration the value of your variable in column (2) is given immediately by the value of b in column (8). In row (9), for instance, $x_1 = 7.5$; in row (13), $x_2 = 7.5$; and so on. For any iteration, then, the profit is simply the sum of *each column (2) value times each column (8) value* for that iteration. The profit for the second iteration, represented by rows (12) to (14), is

$$F_2 = (0 \times 15) + (20 \times 7.5) + (30 \times 7.5) = 375 \text{ or } \$3.75$$

In OR jargon,

$$F_2 = \Sigma(c_{i.2} \cdot b_{i.2})$$

Finding a Place to Start

Once you find a vertex for an initial solution, the simplex tableau

Shows you how to find a more profitable vertex
Tells you when you reached maximum profit
Gives you the product mix that will give you this maximum profit,
and does this even with multidimensional problems too big to visualize

But finding an initial solution presents problems. In the two-dimensional example, you just started at the origin, by setting x_1 and x_2 to zero. You could do the same with a 10-dimensional problem, except that the origin is not always within the feasibility envelope. For a two-dimensional example where the origin is not within the feasibility envelope, suppose you were a two-product feed producer with a contract to provide feed for a dairy cooperative subject to some minimum calorie requirement. Figure 10-5 shows your situation:

A production capacity upper limit of feed No. 1 (vertical line)
A production capacity upper limit of feed No. 2 (horizontal line)
A shipping capacity upper limit for any mix of the two feeds (upper slanting line)

A minimum-calorie *lower* limit for any mix of the two feeds (lower slanting line)

In addition, you will have some objective function line which will dictate the maximum-profit mix.

You can't use the origin for your initial solution, since it isn't in your feasibility envelope. What do you do? (Remember that, although you can see what you're doing in this two-dimensional problem, you need an algorithm that will work when you can't see anything.)

It turns out to be rather simple. In this example, you have two fundamental variables—x_1 and x_2—plus four slack variables (there are four constraints),* but you can't start the simplex tableau for this four-equation, six-variable problem until you find the feasibility envelope. Simply add four more variables, called "dummy variables," and pretend

* The minimum-calorie slack variable has a negative sign, since you must produce this much feed *or more*.

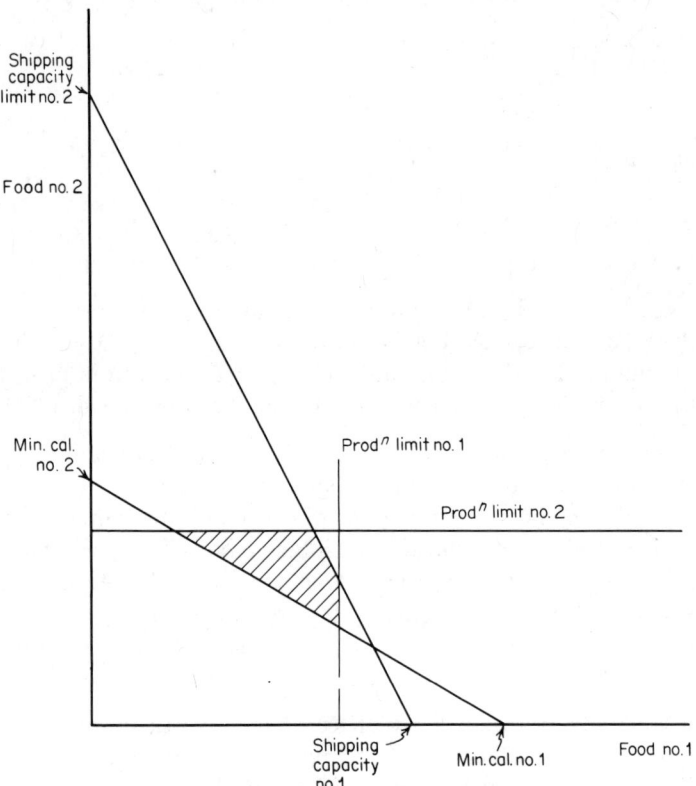

Fig. 10-5

that the five-sided space below and to the left of your feasibility envelope is the area you are interested in. Now you can start at the origin, and work your way from vertex to vertex around this space *until you reach a vertex of the real feasibility envelope*. At this point you have completed what linear programmers call phase I; you can throw away your dummy variables and get on with phase II, the normal simplex solution.

Computer solution of linear programs

On Using Computers

There is a great deal that you do not know about linear programming at this point, but you do know enough to utilize a simple computer program. You know how to express physical constraints (such as the welder's limitations for No. 1 and No. 2 boxes) in equation form. You know how to write an objective function. You know that you must have slack variables (positive if the constraint is an upper limit, negative if it's a lower limit). You know there are such things as dummy variables. You can recognize a solution, expressed in terms of optimum product mix and optimum profit.

And you recognize that, although the preceding discussion has maximized profit, you can use the technique to maximize any other numerical quantity.*

There are many suitable computer programs for solving a linear programming problem. The one you use depends on the available computer and the programming language it uses. Almost any canned program, however, will contain instructions that you can understand and follow with what you now know about the simplex tableau. In the next section, the two-box problem will be solved by a typical computer program.†

The Computer Matrix

Table 10-21 shows the form in which your problem must be set up for using this computer program. Rules must be followed exactly, because the computer won't excuse a single misplaced parenthesis. There are some cautions:
1. Number rows as shown. There is no row 1.
2. Use parenthesis to indicate subscripts: x_1 is written $x(1)$.
3. Write the objective function coefficients across row 0; where a coefficient is zero, omit it. Since this particular program is for

---

* And, by changing all the signs in the objective function, you can handle a problem that seeks to minimize something, such as costs or manpower; the smallest negative number then becomes the "maximum."

† An IBM 7090 computer, using Michigan Algorithm Decoder (MAD) language.

minimizing, and you want to *maximize*, your coefficients have a negative sign (and your final hourly profit will show up with a negative sign).
4. *Every* number in the matrix must have a decimal point, and if there is no number to the right of the decimal point, you must put a 0 there. (The row numbers are for identification only, and they do not have decimal points.)

The Statement Sheet

The computer accepts input from punched cards, each of which holds up to 80 numbers or letters; and it must get them in the right order. You write instructions on a statement sheet which has 80 spaces per line. The keypunch operator will punch a card from each line. The following instructions are written in order, starting with the first line on your statement sheet—which means that they will be in order starting with the first card in your deck.

Computer programs differ, but generally you will have two types of cards:
1. Program cards (the "program deck"), which tell the computer what to do
2. Data cards (the "data deck"), which provide the specific numbers for this particular problem

The *program deck* in this case consists of but 2 cards: the first says who you are and states roughly how big your problem is; the second tells the computer which "canned" program to use.

The *data deck* starts with 4 cards that deal with the mechanics of the setup. If you were using a typewriter instead of a computer, they would be analogous to instructions on setting the margin, the spacing, and so

Table 10-21 A revised simplex algorithm for solving linear programming problems

Row No.	Column no.					
	Actual variables		Slack variables			Constants
	$x(1)$	$x(2)$	$s(1)$	$s(2)$	$s(3)$	b
0	−30.0	−20.0				
2	3.0	3.0	1.0			60.0
3	6.0	2.0		1.0		60.0
4	8.0				1.0	60.0

SOURCE: J. Mehring, R. Disney, and W. A. Spivey, *University of Michigan Utility Programs.*

STATEMENT NO.		FORTRAN STATEMENT		
1-5	6-7	8-30	31-55	56-72
YOUR NAME		UR ACCT NO	SIZE OF UR PROBLEM	
$USE		SIMPLEX		
		↑		
		FOUR		
		ADMINISTRATIVE		
		CARDS		
THREE		EQUATION, TWO UNKNOWN L.P. PROBLEM		
MATRIX				
X(1)		0 -30.0		
X(1)		2 3.0		
X(1)		3 6.0		
X(1)		4 8.0		
X(2)		0 -20.0		
X(2)		2 3.0		
X(2)		3 2.0		
S(1)		2 1.0		
S(2)		3 1.0		
S(3)		4 1.0		
FIRSTB				
		2 60.0		
		3 60.0		
		4 60.0		
		↑		
		FIVE		
		ADMINISTRATIVE		
		CARDS		
		↓		

Fig. 10-6

forth. The description of whatever program you use will contain detailed instructions for these, which you follow blindly. The fifth data card is for a title; give the problem whatever name you like. The name selected in this example is THREE EQUATION, TWO UNKNOWN L.P. PROBLEM.

The next 15 cards contain the specifics of your problem—the characteristics of your three machines and the coefficients in your objective function. They are written in order, going down the first column, then down the second, and so on. If a cell is empty, no card is written for it. The program instructions contain rules pertaining to which of the 80 spaces on the card (or on your statement sheet) to use for which number, and you must follow them exactly.

The first of these 15 cards is MATRIX, and following it you will write the values under your actual variables and slack variables on Table 10-21. (In matrix algebra, these coefficients constitute a "matrix.")

The first 4 cards give the values under $x(1)$, the next 3 cards give the values under $x(2)$, anb so on.

After the matrix has been put on cards, there is a card FIRSTB, following which there are 3 cards for the 3 values of your constant, b.

Your data entries will conclude with 5 final data cards, which you prepare by following the program instructions blindly, just as you did with the first 4 cards of the data deck.

Figure 10-6 shows a statement sheet with these 26 lines (corresponding to 26 cards) filled in. Note that the first 6 and last 5 cards are administrative, or instruction cards; once you have found out what goes on these for the computer you use, you can use them again and again with no further looking up. The set of cards in the middle (corresponding to the middle 15 lines in this example—but the number depends on the size of your problem as given in Table 10-21) is the only thing you will change for each different problem.

You must admit: after you do it once, it's easy.

The Solution

How does the computer tell you the answer? It may clutter it up with a good deal of information you didn't request, because the program you've borrowed may have been written by someone who wanted to know these extra things, but if you look carefully, you will find the values for $x(1)$, $x(2)$, $s(1)$, $s(2)$, $s(3)$, and F.

How long did it take? For this problem, the computer used 2.2 seconds of processing time and 1.8 seconds of execution time, and charged you 20 cents.

11

Input/Output Analysis

CONCEPT

In the late 1930s, the Russian-born economist Wassily Leontief devised a technique for improving market forecasting by assessing the interrelations of one industry with another. His thesis was that, while industry production for final consumption is the effective output of a nation or a region or a company, industry production for *another producing industry* is of great importance in determining the workings of the market. These interindustry transactions tend to be masked by the usual measures of production—they do not appear at all in tabulation of gross national product (GNP), for example—yet no predictions of market potential can be complete without taking them into account.

When automobile production falls, there is an immediate and primary effect on steel production, and this interindustry flow is easy to measure. But the steel industry may provide a market for ceramics, and the ceramic industry a market for industrial chemicals, and the chemical industry a market for laboratory equipment, and so on—and all these secondary industries provide markets for automotive equipment. Thus a cut in automotive production sets in motion ripples that move through the economy and eventually lead to an additional self-induced cut in automotive production. This additional reduction in automotive output initiates a second round of effects, smaller in magnitude, which travels the round and initiates a third round of effects—much as a series of echoes reverberates back and forth across a canyon until the last echo finally dies away.

These principles were understood in a general way, but Leontief made it possible to calculate them specifically. His techniques have been refined and expanded, to the point that input/output now is an accepted tool of market planning for nations, regions, cities, and industrial companies.* It provides a method of calculating the overall effect, direct and indirect, that a change in production level of one industry will have on the demand for products of all the other industries in the economy—including those for which the connection might seem remote or nonexistent.

Construction of the first input/output tables for the United States economy was a monumental task, for everything had to be started from scratch. At the outset, the decision had to be made on how the economy was to be subdivided. If all metal products are grouped into a single sector, for instance, the predicted demand for "metal products" will not be very helpful to a manufacturer who makes metal pencil sharpeners or to a manufacturer who makes metal rails; but if each small subdivision of the economy is treated as a separate sector, the task will be huge. The early work necessarily aggregated the economy into a small number of sectors, but in succeeding years more and more information has been collected, and improved input/output tables are becoming available which refine the information in a way far more useful to specific manufacturers. Today, companies find this technique increasingly valuable for market prediction, and with it they are able to see into the future with more precision than has been possible in the past. Few companies can afford to neglect the potential of input/output analysis for providing detailed and accurate information about the interplay of the industrial market.

In order to understand the concepts involved, it will be helpful to follow through the simplest sort of input/output analysis. The following section will build a table for a community that has been simplified in a highly artificial way.

Imagine a primitive village with just three industries—fishing, agriculture, and weaving. The craftsmen in these industries have learned that they will get more production if they draw to some extent on the products of other industries; thus some of what is produced by one

---

* Celanese Corporation, which produces largely for other industries, finds its input/output model essential for market planning. North American Rockwell uses input/output models to explore the commercial opportunities that "match" its aerospace expertise.

industry will be used up in the production process of another industry and never reach the final consumers (except indirectly, through its contribution to final production of that other industry). These interindustry flows are as follows:

1. The *fishing* industry provides bait for itself, and fertilizer to agriculture (a fish under each hill of corn), but nothing of significance to weaving.
2. The *agriculture* industry provides fish traps to fishing, seed to itself, and fibers to weaving.
3. The *weaving* industry provides nets to fishing, and food bags to agriculture, but nothing of significance to itself.

Suppose that in a given day, each industry must purchase from all industries in the following amounts (of some homogeneous currency—perhaps cattle):

Fishing purchases: 3 units of fishing
                       5 units of agriculture
                       2 units of weaving

Agriculture purchases: 2 units of fishing
                           5 units of agriculture
                           2 units of weaving

Weaving purchases: 11 units of agriculture

These interindustry transactions may be represented on a grid, or matrix, as shown in Fig. 11-1.

These interindustry transactions are incidental to the main business at hand, which is production for final demand. The main production of

		Purchasing Industries		
		Fishing (f)	Agriculture (a)	Weaving (w)
Producing Industries	Fishing (f)	3	2	0
	Agriculture (a)	5	5	11
	Weaving (w)	2	2	0

Fig. 11-1 Interindustry transactions: purchasing industries and producing industries.

the fishing industry goes to the consumer. In a typical day, while 5 units of fish products are being taken for interindustry transactions as shown in Fig. 11-1, 45 units of fish production are going to final demand—or a gross daily production of 50 units. The agriculture industry, while delivering 21 units for interindustry transactions, delivers 19 to final demand for a gross daily production of 40 units. And the weaving industry delivers 4 units for interindustry transactions, 36 to final demand, and 40 to gross daily production. These additional items are shown in Fig. 11-2.

Since input/output is a double-entry system, using some common unit of value such as cattle or dollars, output of each industry must equal its input. If 50 units of fish products are produced, total input to the fish industry must be 50 units. Since the first column of Fig. 11-2 adds to only 10 units of input for the fishing industry (all of which are absorbed within the production process of the industry—so that none come out the other end in the form of consumable fish), another 40 units must be shown as input to balance the system. These 40 units will be in the form of "primary factors" from the fishing sector (in plain language, they start as fish and they are consumed as fish—not as fertilizer or bait), as profit to fishermen, as depreciation on fishing equipment if this is a capital industry, and so on. These items are given the general term "value added." The matrix is completed in this double-entry form for all three industries: inputs are subtotaled beneath each column, outputs are subtotaled to the right of each row, and these subtotals are added horizontally or vertically to the same totals in the bottom right corner of the matrix. Such an interindustry transactions matrix is shown in Fig.

		Purchasing Industries					
		Fishing (f)	Agri-culture (a)	Weaving (w)	Total Industrial use	Final demand (GNP)	Gross Output
Producing Industries	Fishing (f)	3	2	0	5	45	50
	Agri-culture (a)	5	5	11	21	19	40
	Weaving (w)	2	2	0	4	36	40

Fig. 11-2 Interindustry transactions: purchasing industries and producing industries.

INPUT/OUTPUT ANALYSIS

		Purchasing Industries					
		Fishing (f)	Agriculture (a)	Weaving (w)	Total Industrial use	Final demand (GNP)	Gross output
Producing Industries	Fishing (f)	3	2	0	5	45	50
	Agriculture (a)	5	5	11	21	19	40
	Weaving (w)	2	2	0	4	36	40
	Total industry input	10	9	11	30		
	Value added	40	31	29		100	
	Gross input	50	40	40			130

Fig. 11-3 Transactions matrix.

11-3. Note that final demand totals 100; this corresponds to GNP (though GNP is figured for a year, of course, not a day as in the above example), as defined in systems of national accounts. You can see that the total production of the village is larger than this GNP, but the excess is not a usable final product. Note further that value added also totals 100, making it another measure of GNP.

The foregoing constitutes a day in the life of the village, but you do not want to be restricted to daily accounting. You can reduce the figures from Fig. 11-3 to proportions or ratios which will be valid for any time period—will be valid, that is, so long as technology doesn't change and thereby modify the proportions each industry uses of the various factors of production.

In Fig. 11-3, refer to the gross input figures at the bottom of each column. If you express the fraction of this total input which is represented by each of the interindustry inputs, the first column is:

From fishing: $3/50$ or 0.06
From agriculture: $5/50$ or 0.10
From weaving: $2/50$ or 0.04

	Purchasing Industries		
Producing Industries		Fishing (f)	Agri-culture (a)	Weaving (w)
Fishing (f)	0.06	0.05	
Agri-culture (a)	0.10	0.125	0.275
Weaving (w)	0.04	0.05	

Fig. 11-4 Input/output coefficients matrix.

If you do this for all three industries, you have a matrix of input ratios such as that shown in Fig. 11-4. Called the input/output coefficients matrix, it represents the coefficients calculated with respect to the producing industry.

What does this table do for you? It tells you that if you want to produce, say, $100 worth of gross output in agriculture, you must buy $5 worth of fish products, $12.50 worth of farm products, and $5 worth of weaving products as an inescapable part of your production process in farming.

How helpful is this knowledge? Not very—because it ignores the indirect effects that cause requirements to pyramid, and thus is incomplete.

Suppose you have just finished designing a ship, when word comes in that you must increase the size of the deck force by 20 sailors. Each sailor weighs 150 pounds, or perhaps 250 with his sea bag, but when you put this additional 5,000 pounds on board you set a circular process in motion. An increased deck force means more food requirements and hence a few additional hands in the galley (who in turn mean still more food requirements and thus perhaps still another galley hand). The ship must expand a bit to handle these men, which means slightly larger engines to maintain design speed, which means more men in the engine room—by which time the hull is growing to the point that you need even more sailors in the deck force to keep it painted and polished. This process keeps pyramiding, in successively smaller increments, to the point that an additional 150-pound sailor typically adds a ton to the ship's weight.

This same phenomenon obviously occurs in your village industries.

INPUT/OUTPUT ANALYSIS

Suppose you decide to increase the gross output of the fishing industry by $1; the fishing-agriculture coefficient of Fig. 11-4 is .05, so you conclude that agriculture must expand 5 cents to make possible this $1 increase in fish products. But this is only the first round. In order for agriculture to increase in this way, additional inputs will be required, part of which come from the fishing industry. Another part comes from itself, and that part will impose a further interindustry demand on the fishing industry. Your initial decision to increase gross output of fish products by 1 unit will lead you to increase fish products more than 1 unit before this intertwined cause-and-effect process that you set in motion finally dies away.

To make matters worse, this isn't the kind of decision you would make anyway. Your marketing manager is not likely to say: "Let's give gross output an initial boost of $100,000, and see how much we'll have to really boost it before we're done." He's not going to live so dangerously. He's interested in the changes that will occur in each sector if the *ultimate* increase in production is $100,000—and when he says "production," chances are he automatically excludes that part of his own output that goes right back into his production process and never gets outside the plant walls (but he includes the same sort of thing if it goes into the production process of another firm in the same industry). If he produces primarily for consumers, the production he's interested in may be final demand rather than gross output. But whatever his interest, he wants to know market behavior resulting from some *final* volume change, not from initial volume change which can get much larger after the wheels of the interrelated economy stop whirring.

Clearly, the direct coefficients of Fig. 11-4 will not do. You need some different coefficients that can tell you the total effect—direct plus indirect—of a change in some sector, after the entire round robin of interrelationships has taken place. Such coefficients are shown in Fig. 11-5. They show you the fraction by which each using industry (across the top) increases its demand on a producing industry (at left) for 1 unit of increase in final demand of the producing industry. For example, the coefficient in the fishing-agriculture cell (given the designation a_{fa}) of .062 tells you that a $1 increase in *final* demand for fish products necessitates a 6.2-cent increase in *agricultural* demand for fish products. We can total the whole fishing-sector row:

$1.071 increase in fishing-industry uses of fish products*
0.062 increase in farming-industry uses of fish products
<u>0.017</u> increase in weaving-industry uses of fish products
$1.150 increase in total industry requirements for fish products

* Includes the $1 increase in final demand.

	Purchasing Industries		
	Fishing (f)	Agriculture (a)	Weaving (w)
Producing Industries — Fishing (f)	1.071	0.062	0.017
Producing Industries — Agriculture (a)	0.138	1.169	0.322
Producing Industries — Weaving (w)	0.050	0.061	1.016

Fig. 11-5 Inverse matrix.

This industry requirement, including the $1 increase in final demand, gives a $1.150 increase in gross output.

(An interim point should be noted here. The weaving industry has no direct requirements for fish products, as you see from Fig. 11-1, but it has requirements for farm products, and agriculture has requirements for fish products; changes in weaving output therefore clearly affect the fishing industry—and vice versa.)

Note how different the ratios are from those of Fig. 11-4, and how misled you would be if you based your conclusions about the effect of changes in one industry sector on what you notice in Fig. 11-2. The latter suggests that final demand is $45/50$, or 90 percent, of gross output—and so it is for the exact product mix of that table. But if you increase the production of a single industry as you have just done, without any modification in final demand of other industries, the proportions change.

How is Fig. 11-5 constructed? Computation of the coefficients in Fig. 11-4 was quite direct, but Fig. 11-5 is a different story. Figure 11-5 is an "inverse matrix," and to understand how it is constructed we should review the basic concepts of matrix manipulation.

MATRICES IN EVERYDAY USE

A matrix is simply a rectangular array of numbers, which presumably has some meaning for you. Table 11-1 shows examples of matrices:

A rectangular matrix   Dimensions 2×3
A column vector   Dimensions 4×1
A row vector   Dimensions 1×2
A square matrix   Dimensions 3×3

INPUT/OUTPUT ANALYSIS

Table 11-1

Rectangular matrix	Column vector
$\begin{bmatrix} 2 & 3 & -4 \\ 7 & -2 & 5 \end{bmatrix}$	$\begin{bmatrix} 6 \\ 11 \\ 5 \\ 9 \end{bmatrix}$
2×3	4×1
Row vector	Square matrix
$[5 \quad 3]$	$\begin{bmatrix} 1 & 0 & 1 \\ 3 & 1 & 0 \\ 2 & 1 & 4 \end{bmatrix}$
1×2	3×3

Generalized matrix

$$A = \begin{bmatrix} a_{11} & a_{12} & a_{13} \\ a_{21} & a_{22} & a_{23} \\ a_{31} & a_{32} & a_{33} \end{bmatrix}$$

Each number in a matrix is called an "element," and in standard numbering systems it bears the number of the row followed by the number of the column. A generalized matrix uses a lower-case letter for each element, with subscripts showing what row and column it is in. Shorthand nomenclature for such a generalized matrix is simply to call it by a single capital letter corresponding to the lower-case letter used for each of its elements. Thus, in Table 11-1 the square 3×3 matrix whose elements are a_{11}, a_{12}, etc., is identified simply by the letter A. The letter doesn't tell you the dimensions of the matrix it represents, but generally there are other clues that take care of this.

Matrix algebraists are fond of referring to "the a_{ij}th element." This simply means an element in the ith row and in the jth column, where no specific row or column is specified—and indeed, where the intention may be to have i or j take successive values corresponding to several or all of the rows or columns.

The shipping clerk in a steel mill may use matrices constantly, without ever realizing it. Suppose that in a given period his mill, Mill S, turns out the following tonnage of various products:

Rods	Shapes	Plates	Total
20	8	32	60

Suppose that he is interested in the distribution of this tonnage to broad sectors of his market, segregated by product. He could tabulate this conveniently by using a matrix such as that shown at the top of Table 11-2. In another mill, Mill T, the product/market matrix might look as shown at the bottom of Table 11-2.

If an industry spokesman wanted to tabulate the combined output of Mills S and T, he would add the two matrices by simply adding corresponding elements one at a time. Combined rod production for industrial users would be $4 + 2$. If you were using a letter designator for each element in Mill S of s_{ij}, and in Mill T of t_{ij}, then the combined rod production for industrial users would be $s_{ij} + t_{ij}$. If you wanted to indicate that you carried through this addition 9 times, for the 9 elements, you could write simply $C = S + T$, where C is the combined product/market matrix for the two mills (see Table 11-3).

So much for matrix addition. Matrix multiplication is equally straightforward. Suppose the selling price per ton were:

Rods: $300
Shapes: 400
Plates: 240

Table 11-2

	Mill S			
	Rods	Shapes	Plates	Total by outputs
Industrial	4	5	16	25
Building	16	2	6	24
Consumers	0	1	10	11
Total by inputs	20	3	32	60

	Mill T			
	Rods	Shapes	Plates	Total by outputs
Industrial	2	11	8	21
Building	9	3	0	12
Consumers	1	0	7	8
Total by inputs	12	14	15	41

Table 11-3

$$\begin{bmatrix} 4+2 & 5+11 & 16+8 \\ 16+9 & 2+3 & 6+0 \\ 0+1 & 1+0 & 10+7 \end{bmatrix} = \begin{bmatrix} 6 & 16 & 24 \\ 25 & 5 & 6 \\ 1 & 1 & 17 \end{bmatrix}$$

$$C = S + T = \begin{bmatrix} s_{11} & s_{12} & s_{13} \\ s_{21} & s_{22} & s_{23} \\ s_{31} & s_{32} & s_{33} \end{bmatrix} + \begin{bmatrix} t_{11} & t_{12} & t_{13} \\ t_{21} & t_{22} & t_{23} \\ t_{31} & t_{32} & t_{33} \end{bmatrix}$$

$$C = \begin{bmatrix} s_{11} + t_{11} & s_{12} + t_{12} & s_{13} + t_{13} \\ s_{21} + t_{21} & s_{22} + t_{22} & s_{23} + t_{23} \\ s_{31} + t_{31} & s_{32} + t_{32} & s_{33} + t_{33} \end{bmatrix}$$

It is easy to see that the receipts at Mill T can be calculated by multiplying each product tonnage by the price per ton of that product:

For rods:	12 × $300 =	$ 3,600
For shapes:	14 × 400 =	5,600
For plates:	15 × 240 =	3,600
Total:		$12,800

If you are willing to write these two sets of numbers in a particular arrangement, and multiply in accordance with a set sequence, you can do the same calculation as matrix multiplication. What you must do is write the tons of production for each product as a *row vector*, and the price per ton for each product as a *column vector*, and then remember that *the first item in the row of the first matrix multiplies only by the first item in the column of the second matrix.* Any other sequence in matrix multiplication is a prohibited operation (as well it should be, since multiplying tons of plates by price per ton of rods makes no sense). Table 11-4 shows this operation at the top of the table. While you are about it, you can as easily do the same calculation for Mill S, since the prices apply equally to both mills. The combined operation for both mills is shown at the bottom of Table 11-4.

Perhaps some rules of matrix multiplication now are apparent. First is the fact that, as you go across the row in the left-hand matrix, you go down the column in the right-hand matrix, and you can multiply only corresponding elements together. Second, as you go across a row and down a column in this way, you add all the products of one such traverse together (production of Mill S is the sum of $6,000 + $3,200 + $7,680). Third, in order for you to have an equal number of things to multiply together, the number of columns in the left matrix must equal the number of rows in the right matrix. (This is the equality marked "test" with an arrow, at the bottom of Table 11-4.)

Table 11-4

Mill T

$$\begin{array}{ccc} R & S & P \\ [12 & 14 & 15] \end{array} \times \begin{array}{c} R \\ S \\ P \end{array} \begin{bmatrix} 300 \\ 400 \\ 240 \end{bmatrix} = [3{,}600 + 5{,}600 + 3{,}600] = [12{,}800]$$

Mills S and T

$$\begin{array}{ccc} R & S & P \\ \begin{bmatrix} 20 & 8 & 32 \\ 12 & 14 & 15 \end{bmatrix} \end{array} \times \begin{array}{c} R \\ S \\ P \end{array} \begin{bmatrix} 300 \\ 400 \\ 240 \end{bmatrix} = \begin{bmatrix} 6{,}000 + 3{,}200 + 7{,}680 \\ 3{,}600 + 5{,}600 + 3{,}600 \end{bmatrix} = \begin{bmatrix} 16{,}880 \\ 12{,}800 \end{bmatrix}$$

Generalizing

$$A = \begin{bmatrix} a_{11} & a_{12} & a_{13} \\ a_{21} & a_{22} & a_{23} \end{bmatrix} \times \begin{bmatrix} b_1 \\ b_2 \\ b_3 \end{bmatrix} = \begin{bmatrix} a_{11}b_1 + a_{12}b_2 + a_{13}b_3 \\ a_{21}b_1 + a_{22}b_2 + a_{23}b_3 \end{bmatrix}$$

Note that $AB \neq BA$.

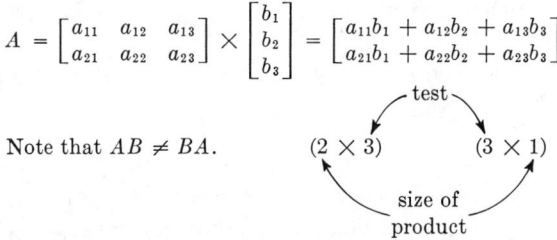

$(2 \times 3) \quad (3 \times 1)$

test

size of product

MATRIX OPERATIONS

Hold off on matrix division for a bit, and consider a simple problem in matrix multiplication. Suppose you had the following situation: a square 2×2 matrix is to be multiplied by a 2×1 column vector. You know the multiplication is possible, because the two inner values are

Table 11-5

$$\begin{bmatrix} 2 & 1 \\ 3 & 2 \end{bmatrix} \times \begin{bmatrix} x_1 \\ x_2 \end{bmatrix} = \begin{bmatrix} 5 \\ 8 \end{bmatrix}$$

Generalizing

$$\begin{bmatrix} a_{11} & a_{12} \\ a_{21} & a_{22} \end{bmatrix} \times \begin{bmatrix} x_1 \\ x_2 \end{bmatrix} = \begin{bmatrix} b_1 \\ b_2 \end{bmatrix}$$

Or in matrix notation:
$\quad AX = B$

INPUT/OUTPUT ANALYSIS

equal—each is 2. You know the dimensions of the product—it takes the dimensions established by the two outer values, or 2×1; so it is another column vector. The values of these matrices are given in Table 11-5.

If you carry out this matrix multiplication, you get:

$$2x_1 + x_2 = 5$$
$$3x_1 + 2x_2 = 8$$

This is a simple set of simultaneous equations, easy to solve for the values of x_1 and x_2 by several methods. You can use simple substitution, or you can multiply the top equation by (-2) and add to eliminate x_2, or you can use determinants if you recall them. Or you can solve for X (that is, for x_1 and x_2 written as a column vector) by *inverting the matrix*.

Many matrix operations correspond to algebraic operations. If you were solving a simple algebraic expression such as:

$$ax = b$$

you would simply write it as:

$$x = \frac{b}{a}$$

which is another way of saying

$$x = b\left(\frac{1}{a}\right)$$

which can be written in still another form as

$$x = b(a)^{-1}$$

If you write the previous problem in matrix multiplication in its generalized form, using coefficients in the form of letters with subscripts, it would appear as shown in Table 11-5. The matrix shorthand for this same notation, using capital letters, is shown beneath it. And just as you can write the *reciprocal* of a in the algebraic expression above, by using the convention of $(a)^{-1}$, so you can write the *inverse* of the matrix A in our matrix problem, using the same convention: A^{-1}. Furthermore, when used in this form it accomplishes the same mathematical result. In the algebraic example above, when you multiplied b by the reciprocal of a you actually solved in one step for x. In matrix algebra, when you accomplish the matrix multiplication operation of $A^{-1}B$ (in this order; the other order won't work), *you solve in one step for every x in your set of equations;* in other words, you solve for the column vector, X. It is hardly worth doing for two equations with two unknowns—that is, for an A matrix of dimensions 2×2; but when you get an A matrix of

Table 11-6

$$\begin{bmatrix} 2 & 1 \\ 3 & 2 \end{bmatrix} \times \begin{bmatrix} i_{11} & i_{12} \\ i_{21} & i_{22} \end{bmatrix} = \begin{bmatrix} 1 & 0 \\ 0 & 1 \end{bmatrix}$$

$$= \begin{bmatrix} 2i_{11} + 1i_{21} & 2i_{12} + 1i_{22} \\ 3i_{11} + 2i_{21} & 3i_{12} + 2i_{22} \end{bmatrix}$$

Equating the products element by element:

$$\begin{array}{c} 2i_{11} + i_{21} = 1 \\ 3i_{11} + 2i_{21} = 0 \end{array} \quad \text{and} \quad \begin{array}{c} 2i_{12} + i_{22} = 0 \\ 3i_{12} + 2i_{22} = 1 \end{array}$$

Solving the above pairs of equations:

$$A^{-1} = \begin{bmatrix} i_{11} & i_{12} \\ i_{21} & i_{22} \end{bmatrix} = \begin{bmatrix} 2 & -1 \\ -3 & 2 \end{bmatrix}$$

Using the inverse matrix to solve for x_1 and x_2:

$$X = A^{-1}B = \begin{bmatrix} 2 & -1 \\ -3 & 2 \end{bmatrix} \times \begin{bmatrix} 5 \\ 8 \end{bmatrix} = \begin{bmatrix} 10 - 8 \\ -15 + 16 \end{bmatrix} = \begin{bmatrix} 2 \\ 1 \end{bmatrix} \begin{matrix} (x_1) \\ (x_2) \end{matrix}$$

---

dimensions 25 × 25, or 106 × 106, no other way of reaching a solution is feasible at any acceptable cost.

How do you invert a matrix? Conceptually it is very simple. Just as in algebra we know that:

$$a(a)^{-1} = 1$$

so we know in matrix algebra that:

$$AA^{-1} = I$$

where I is something known as an identity matrix. It has the characteristic that anything multiplied by it is unchanged, and thus it occupies the same place in matrix algebra that 1 occupies in conventional algebra. Regardless of its size, it always looks the same: every element in it is 0 except those elements on the principal diagonal (the one running from upper left to lower right corners), and all the latter are 1. Thus elements a_{11}, a_{22}, a_{44}, and so on are 1, all others are 0. You can therefore solve for A^{-1} by asking yourself, "What is it which, when multiplied by A, gives I?"

Table 11-6 carries through this method of inverting the A matrix of Table 11-5. Having got the inverse matrix, or A^{-1}, it solves for the variable matrix X by the simple formula described above:

$$X = A^{-1}B$$

and gets the values shown at the bottom of the table:

$$x_1 = 2$$
$$x_2 = 1$$

INPUT/OUTPUT ANALYSIS

The only thing wrong with this method is that it involved (as you can see in Table 11-6) two problems in simultaneous equations, each of which was as extensive as the entire original problem. This method, therefore, is not the one you actually use except in special cases. It is beyond the scope of this discussion to outline actual methods of inverting a matrix, which are done by computer in any case; but it should be clear why an inverted matrix is useful in solving large problems of linear equations. The general forms of matrix notation should be understood by this point, so that you will be able to follow the operations of the next section dealing with computation of indirect coefficients.

MATRIX INVERSION IN INPUT/OUTPUT ANALYSIS

In order to see how the matrix inversion process is used to compute the direct plus indirect effects, it will be necessary to go through a few steps in matrices. If you refer to Fig. 11-2, you see actual numbers inserted in the matrix transactions. Table 11-7 contains the same entries (excluding the "Total industry output" subtotal), but expressed in

Table 11-7

	Purchasing industries			Final demand	Gross output
	f	a	w		
f	x_{11}	$+x_{12}$	$+x_{13}$	$+y_1 = x_1$	
a	x_{21}	$+x_{22}$	$+x_{23}$	$+y_2 = x_2$	
w	x_{31}	$+x_{32}$	$+x_{33}$	$+y_3 = x_3$	

Simplifying to a 2 × 2 matrix for brevity makes this

$$x_{11} + x_{12} + y_1 = x_1$$
$$x_{21} + x_{22} + y_2 = x_2$$

Which can be written, without change in value, as follows:

$$\left(\frac{x_{11}}{x_1}\right) x_1 + \left(\frac{x_{12}}{x_2}\right) x_2 + y_1 = x_1$$
$$\left(\frac{x_{21}}{x_1}\right) x_1 + \left(\frac{x_{22}}{x_2}\right) x_2 + y_2 = x_2$$

generalized form as variables. Within the 3 × 3 matrix, x_{11} is the interindustry transfer from sector 1 (fishing) to sector 1; x_{12} is the interindustry transfer from sector 1 to sector 2 (agriculture); and x_{13} is the interindustry transfer from sector 1 to sector 3 (weaving). Outside the matrix, y_1 is the final demand for sector 1, x_1 is the gross output for sector 1, and so forth for the remaining producing sectors. The transactions of each producing sector can be expressed as an equation as shown.

These equations are shown below the matrix of Table 11-7 (reduced to dimensions 2 × 2 in the interest of brevity—though the treatment for any size of matrix would be the same):

$$x_{11} + x_{12} + y_1 = x_1$$
$$x_{21} + x_{22} + y_2 = x_2$$

If you divide each of the first terms by x_1/x_1 (or multiply, if you prefer), and each of the second terms by x_2/x_2, you have changed nothing, but the equations are in the form shown at the bottom of Table 11-7.

The variables in parentheses are the direct coefficients which you saw in Fig. 11-4. For example, x_{12}/x_1 is the amount of sector 1 output used in sector 2's production process, expressed as a proportion of the gross output (or input) of sector 1. If you were using a to signify the value of each direct coefficient in Fig. 11-4, this would be a_{12}. If you make this substitution, the equations will look as they do at the top of Table 11-8:

$$a_{11}x_1 + a_{12}x_2 + y_1 = x_1$$
$$a_{21}x_2 + a_{22}x_2 + y_2 = x_2$$

Table 11-8 The matrix of coefficients for direct and indirect requirements per unit of final demand.

$x_1 = a_{11}x_1 + a_{12}x_2 + y_1$
$x_2 = a_{21}x_1 + a_{22}x_2 + y_2$

In matrix form,

$$\begin{bmatrix} x_1 \\ x_2 \end{bmatrix} = \begin{bmatrix} a_{11} & a_{12} \\ a_{21} & a_{22} \end{bmatrix} \times \begin{bmatrix} x_1 \\ x_2 \end{bmatrix} = \begin{bmatrix} y_1 \\ y_2 \end{bmatrix}$$
$$X = AX + Y$$
$$X - AX = Y$$
$$(I - A)X = Y$$
$$X = (I - A)^{-1}Y$$

If we call this "inverse matrix" by A', then

$$\begin{bmatrix} x_1 \\ x_2 \end{bmatrix} = \begin{bmatrix} a'_{11} & a'_{12} \\ a'_{21} & a'_{22} \end{bmatrix} \times \begin{bmatrix} y_1 \\ y_2 \end{bmatrix}$$

INPUT/OUTPUT ANALYSIS

This can be expressed as a matrix multiplication in long form, as shown in the table; and this in turn can be shown in matrix notation:

$$X = AX + Y$$

If we rearrange this, collecting the X on the left side of the equation, we can solve for X by making use of the matrix inverse operation, as follows:

$$X = (I - A)^{-1}Y$$

This equation is a very interesting one. The inverse matrix, $(I - A)^{-1}$, which is of the same dimensions as our original direct coefficients matrix, A (the one shown in Fig. 11-4), is the matrix of a new set of coefficients constituting the direct and indirect requirements per unit of final demand. Such a set of coefficients is shown in the matrix of Fig. 11-5. If you have such a set of coefficients, all you need to do is post-multiply* by the column vector of final demand, and you come out with the required column vector of gross output. All you need now is a method for calculating this inverse matrix, given that you know the technological coefficients of the economy with which you are dealing (that is, the matrix A, or the values of Fig. 11-4).

There are several methods of matrix inversion, all very time-consuming if you are doing them by hand. The preferred one for this particular purpose is an infinite series approximation technique, described below.

In conventional algebra, if you want to evaluate the expression $(1 - a)^{-1}$ which is the same as $1/(1 - a)$, you can simply add together all the values of a^n, as n takes in succession every integral value from zero to infinity (provided that a has some value between 0 and 1). Thus: $1/(1 - \frac{1}{3})$ is equal to the limit of $(\frac{1}{3})^0 + (\frac{1}{3})^1 + (\frac{1}{3})^2 + \cdots$. This series works the same way in matrix manipulation. That is,

$$(I - A)^{-1} = I + A + A^2 + A^3 + A^4 + \cdots$$

Therefore, combining this with your previous equation, and as shown in Table 11-9, the value of the matrix X is obtained by post-multiplying each term in the above series by the matrix Y.

This would promise to be a long process. For your purposes, however, sufficient accuracy is achieved for terms past the third power of

---

* The term "post-multiply" needs explanation. In regular arithmetic, if you are asked to multiply a number by 3 (perhaps the number 17), you write 17 × 3—but you could as accurately have written 3 × 17. This is not true of matrix arithmetic, as was pointed out in Table 11-4. Thus it is necessary to use the term "post-multiply" by a number, to indicate that the number comes second in the multiplication rather than first.

Table 11-9 Formula for matrix inversion

$$(I - A)^{-1} = I + A + A^2 + A^3 + \cdots$$
$$X = IY + AY + A^2Y + A^3Y + \cdots$$
$$X \doteq IY + AY + A^2Y + A^3Y + \frac{(A^3)^2}{A^2 - A^3} Y$$

the A matrix if we approximate them with the following function:

$$\frac{(A^3)^2}{A^2 - A^3}$$

This approximate formula for evaluating your problem is shown at the bottom of Table 11-9.

This series expansion is more than a convenient way to invert the matrix; it gives specific expression to the infinite round of increases that is triggered by making one change or more in final demand vectors:

The first term, IY, is the direct production needed for final demand.

The second term, AY, is the production needed to provide inputs into the direct production for final demand.

The third term, A^2Y, is the production needed to provide inputs into the above inputs.

The fourth term, A^3Y, is the production needed to provide inputs to the inputs to the inputs.

All the remaining decreasing ripples in the interindustry process are taken care of accurately enough by the final approximate term.

COMPUTATION OF INDIRECT COEFFICIENTS

It is interesting to see how these successive values behave as they dwindle away to zero. Take as an example the upper left corner of the village-industry matrix in Fig. 11-3. The direct requirements coefficient shown in the fishing-fishing cell of Fig. 11-4 is .06. If final demand for this sector (y_1 of the Y vector) is 1.0, the successive terms of the above-described expansion are shown below for this cell:

First round (IY)	1.000000
Second round (AY)	.060000
Third round (A^2Y)	.008600
Fourth round (A^3Y)	.001991
Remaining round	.000600
Gross output	1.071190

INPUT/OUTPUT ANALYSIS

Since this gross output is for a unit value of final demand, it is a coefficient which is multiplied by the actual final demand.

Table 11-10 shows the results of this computation, carried out for the entire matrix. The totals for the cells correspond to the values for the inverse matrix, Fig. 11-5.

How is Fig. 11-5 (or Table 11-10) used? Just as the matrix multiplication operation indicates:

$$\frac{\text{Gross output}}{\text{column vector}} = \frac{\text{inverse}}{\text{matrix}} \times \frac{\text{final demand}}{\text{column vector}}$$

$$X = (I - A)^{-1} \times Y$$

The final demand specified for fishing is 45 units. Look at the first (fishing) column of Table 11-10; reading down you see the fractions of 45 units in each cell, which must be multiplied by 45 to give the value that would result from the matrix multiplication.

The matrix multiplication is shown in Table 11-11. Note that if you subtract from the gross production figures in the expanded product matrix of Table 11-11 the final demand figures (that is, 48.204 − 45, 22.210 − 19, and 36.593 − 36), you are left with approximately the same value for interindustry transactions as you had initially in Fig. 11-2.

Table 11-10 Computation of direct and indirect coefficients

1.000000	.000000	.000000
.060000	.050000	.000000
.008600	.009250	.013750
.001991	.002274	.002544
.000600	.000741	.000577
1.071190	.062265	.016871
.000000	1.000000	.000000
.100000	.125000	.275000
.029500	.034375	.034375
.006583	.007491	.009453
.001891	.002087	.003586
.137973	1.168953	.322414
.000000	.000000	1.000000
.040000	.050000	.000000
.007400	.008250	.013750
.001819	.002089	.002269
.000593	.000708	.000448
.049812	.061047	1.016467

Table 11-11 Estimated gross production

$$\begin{bmatrix} 1.071 & 0.062 & 0.017 \\ 0.138 & 1.169 & 0.322 \\ 0.050 & 0.061 & 1.016 \end{bmatrix} \times \begin{bmatrix} 45 \\ 19 \\ 36 \end{bmatrix} = \begin{bmatrix} 48.204 & 1.183 & 0.607 \\ 6.209 & 22.210 & 11.607 \\ 2.242 & 1.160 & 36.593 \end{bmatrix} = \begin{bmatrix} 49.994 \\ 40.026 \\ 39.994 \end{bmatrix} \begin{matrix} (50) \\ (40) \\ (40) \end{matrix}$$

Subtracting final demand, and comparing with Fig. 11-2 interindustry values:

$$\begin{bmatrix} 3.20 & 1.18 & 0.61 \\ 6.21 & 3.21 & 11.61 \\ 2.24 & 1.16 & 0.59 \end{bmatrix} \text{ vs. } \begin{bmatrix} 3 & 2 & 0 \\ 5 & 5 & 11 \\ 2 & 2 & 0 \end{bmatrix}$$

The two sets of values are compared at the bottom of Table 11-11. They are approximately the same, but there are significant differences. When you looked at the top row of Fig. 11-2, you saw that fishing absorbed 3 units for its own production, agriculture absorbed 2 units of fish products, and weaving absorbed no fish production. Later, the fact was discussed that some of this absorbing of fishing output was to make it possible for that sector to contribute from its own output to fishing output. It is instructive to look at the matrix at the bottom of Table 11-11 on the left as the number of units a sector absorbs for that part of its production that contributes to its own final demand. As such, these are not actual transfers of production between industries (the bottom right matrix shows actual transfers); perhaps they might be thought of as net transfers that occur after all the round-robin interdependencies have taken effect.

USE OF INPUT/OUTPUT TABLES

Figure 11-6 shows an excerpt from an actual input/output table for the national economy.* The full table contains 106 different sectors, of which only 3 sectors are shown. Within the matrix proper there are two entries in each cell: the upper one is the volume of direct sales, and the lower one is the volume of direct plus indirect sales (rounded to the nearest 10). All figures are in millions of dollars. The three sectors selected here have only minor direct interdependencies: thus, meat products as a sector uses only about $10 million directly from the printing and publishing sector—but uses $440 million directly and indirectly. This table corresponds to Fig. 11-3 of the simplified example, with the figures in parentheses corresponding to the bottom left section of Table 11-11.

Input/output tables are used by the national government to predict growth of the United States economy by separate sectors, and to assess

---

* This particular table is the "Fortune–C.E.I.R." table for 1966, built by using mathematical techniques to update a table for an earlier year.

the effect on selective growth of various governmental policies. They are used by regions of the country (states or major cities) to identify industrial imbalances and guide the encouragement of new industries in the necessary areas. They are used by individual companies to identify promising areas for expansion or diversification, to measure the effect on future sales of changes in the output of other industries, or to assess the attainable volume for their products under various external conditions. Their chief merit for prediction purposes lies in their ability to show the separate effects of all significant factors, and thereby to permit planners to react more effectively and promptly to changes in part of the total environment, whereas previous prediction techniques tend to give overall market predictions based on lumping all applicable factors into one.

If any significant part of your company's output goes into input of the products of other companies, input/output analysis will play an important part in your market predictions in the future. It is well to become acquainted with it now.

	Meat products (20)	Printing and publishing (40)	Farm machinery (60)	Total industry use	Final demand (GNP)	Gross output
Meat products (20)	3260 (26570)	-- (10)	-- (--)	5060	23470	28530
Printing and publishing (40)	10 (440)	2400 (5430)	-- (50)	16020	4710	20730
Farm machinery (60)	-- (80)	-- (--)	180 (3400)	1210	3270	4480
Total industry input	24100	10610	2720	733,350		
Value added	4430	10120	1760		739,590 GNP	
						1,472,930

Fig. 11-6 Extract from "Fortune–C.E.I.R." table, 1966 (106 × 106).

epilogue

The Future

Operations research and related techniques have taken such a strong hold in business and governmental management that virtually every student coming from university programs in public or business administration considers himself unprepared without grounding in these methods. As federal, state, and local governments turn increasingly to private industry for solutions to public problems, it is taken practically for granted (and often stipulated specifically) that these techniques will be employed. Indeed, government's belief that private corporations are skilled practitioners of the "systems approach"—though often unjustified—has had much to do with the growth in private delegation of public tasks.

One reason for the governmental swing to quantitative techniques is the very scope of governmental problems. Too vast and complex to be comprehended by the unaided managerial eye, they have seemed totally incapable of solution until simplifying techniques could order the issues and make them comprehensible. Even the most decisive manager cannot exercise his powers of choice until some glimmer of rational alternatives looms through the chaos of unordered and conflicting information. Operations research tools have offered real hope of creating sufficient order to permit government administrators to function in their complex environments, and the result has been growing awareness that these tools are essential to efficient government management.

The need in business is scarcely less acute, as swift changes in communications, transportation, technology, and markets have brought complexity even to relatively small enterprises. The businessman is finding customers, suppliers, competitors, and new management recruits all speaking in quantitative terms, and he must do the same. Competitors can move into new areas suggested by systems-analysis studies, and can obtain strong market positions before an executive who uses more traditional measures can recognize that opportunities are unfolding. When a strong company suddenly comes on hard times, the problem may not be just one of unkind fate, but more one of failure to recognize trends that were there for careful analysis to uncover.

It is inexcusable for a business to get in trouble because it does not make use of quantitative techniques. Managers who do not know the value of these methods, who do not have an informed notion of where and how they can be used as well as how to interpret their results, are not fully prepared for the environment of modern business. It is not enough to understand that operations research can be useful to business firms. Managers also have an obligation to *use* it: to seek out applications, to be forever alert to areas which will benefit from its application, and to make the attitude of analysis which it represents a fundamental approach to their everyday problems.

Index

Index

Break-even analysis, 97
Business indicators, 54
    analysis, 59
    categories, 54
    selection, 58

Capital expenditures, 92
    break-even analysis, 97
    check list for capital projects, 99
    costs, 94
    discounted cash-flow, 101
    multiproduct analysis, 98
    payback period, 100
    present worth, 101
    return on investment, 100
    revenue, 95
    types of goods, 94
    uncertainty, 93, 101
Computers:
    in banking, 2
    in linear programming, 201
Correlation (*see* Regression analysis)
Cost-time trade-off (networks), 127

Duckworth, E.:
    Operational Research, 167

Econometrics, 7
Evolutionary operations, 36
Expected value:
    of a gamble, 82
    and "rationality," 83
    relation to utility, 83
    of stock, 84
Exponential distribution:
    for service rates, 168

Future, the 226

Goetz, Billy E:
    Quantitative Methods, 103

Incremental analysis, 85
    marginal profit method, 87
Indifference curve, 7

Information:
  economics of, 15
  value of, 89
Input/output analysis, 205
  concept of, 205
  direct coefficients matrix, 210, 220
  indirect coefficients matrix, 212, 223
  interindustry transactions matrix, 209
  matrix inversion in, 219
  use of, 224
Inspecting (*see* Quality control)
Inventory management, 68
  basic inventory problem, 70
  demand analysis, 81
  economic order quantity, 72
  economics of, 69
  incremental analysis, 85
  optimum level, 71
  with production and use, 74
  safety stocks, 78
  total cost, 77
  types of cost, 70

Lanchester, Frederick William:
  N-square law, 3
  Nelson's strategy according to, 3
Linear programming (*see* Programming)
Locational analysis, 112
  franchise area, 115
  industrial location principles, 119
  interception, 118
  interchange, 117
  retail location principles, 114
  retail volume, elements of, 114–115
  shared area, 116
  trading area, 118
Loss tables:
  in assignment method, 175

McNamara, Robert:
  military systems analysis, introduction by, 4
Marginal analysis, 7
Market studies, 112
  concept of, 120
  procedure, 121
  retail volume, elements of, 114

Matrices, 212
  in everyday use, 212
  identity matrix, 218
  in input/output analysis, 219
  inverse matrix, 217
  operations in, 216
Moody, Leland A:
  Management Sciences in Accounting, 6
Morgenstern, Oskar:
  Theory of Games and Economic Behavior, 103
Multiproduct analysis, 98

Network management:
  algorithms, 135
  concept of management control, 124
  cost-time trade-off, 127, 130
  critical path method, 125, 140
  diagramming, 126
  fundamentals of, 125
  input, 126
  matrix solution, 143
  normal and crash times, 128
  PERT, 125
  role of, 123
  slack, 140, 146
  total cost analysis, 134
  workload smoothing, 148
Normal probability distribution:
  concept of, 16
  curve, 18
  table of, 37

Operations research:
  background of, 2
  basic concepts of, 6
  definition of, 2
  elements of, 7
  evolution of, 4
  Methods of Operations Research (Morse & Kimball), 4
Opportunity loss:
  of stockouts, 79

Pascal's triangle, 17
Poisson distribution:
  for arrival times, 167

INDEX

Prediction equations (*see* Regression analysis)
Preposterior analysis, 36
Profitability analysis, 94
Programming, 173
  assignment method, 174
    algorithm, 178
    loss tables, 175
    step-by-step solution, 174
    summary, 179
  linear programming, 185
    computer solution of, 201
    constraints, 194, 200
    determinate linear systems, 185
    graphical representation, 191, 192
    indeterminate linear systems, 190
    objective function, 192
    simplex method, 193
    simplex tableau, 197
  need for, 173
  transportation method, 180
    improving the solution, 182
    initial solution, 181

Quality control, 31
  optimum sample size, 32
Questionnaires (*see* Surveys)
Queues, 154
  build-up of, 158, 166
  characteristics of, 154
  effect of service ratio on, 164
  length of line, 165
  simulation of, 155
  start of, 162

Regional economic balance, 113
Regression analysis:
  accuracy of, 50
  confidence of, 50
  least-squares fit, 49
  spurious correlation, 51
  stepwise multiple regression, 62, 67

Replacement analysis, 103
  equipment that deteriorates, 104
  equipment that fails, 109
  failure probability, 111
  optimum replacement time, 106
  survival probability curve, 110

Sampling:
  combined with judgment, 29
  distribution of sample means, 24
  effect of sample size, 20
  reliability, 29
  sequential, 33
  versus taking a census, 21
  whether to sample or not, 34
Seasonal adjustment:
  factors, 42
  of profit figures, 40
Simulation, 169
  using random numbers, 171
  of waiting lines, 171
Standard gamble, 7
Surveys:
  design, 24
  interviewer error, 26
  measurement problems, 27
  pilot study, 25
  pitfalls, 24, 26
  questionnaire errors, 25
  response errors, 26

Von Naumann, John:
  Theory of Games and Economic Behavior, 103

Waiting lines, 150
  break-even analysis, 152
  dual nature of, 152
  examples, 153
  waiters and servers, 151

T
57.6
B7

MAR 15 1973